Springer Undergraduate Mathematics Series

Springer
London
Berlin
Heidelberg
New York
Hong Kong
Milan
Paris
Tokyo

Advisory Board

Other books in this series

A First Course in Discrete Mathematics *I. Anderson*
Analytic Methods for Partial Differential Equations *G. Evans, J. Blackledge, P. Yardley*
Applied Geometry for Computer Graphics and CAD *D. Marsh*
Basic Linear Algebra *T.S. Blyth and E.F. Robertson*
Basic Stochastic Processes *Z. Brzeźniak and T. Zastawniak*
Elementary Differential Geometry *A. Pressley*
Elementary Number Theory *G.A. Jones and J.M. Jones*
Elements of Abstract Analysis *M. Ó Searcóid*
Elements of Logic via Numbers and Sets *D.L. Johnson*
Further Linear Algebra *T.S. Blyth and E.F. Robertson*
Geometry *R. Fenn*
Groups, Rings and Fields *D.A.R. Wallace*
Hyperbolic Geometry *J.W. Anderson*
Information and Coding Theory *G.A. Jones and J.M. Jones*
Introduction to Laplace Transforms and Fourier Series *P.P.G. Dyke*
Introduction to Ring Theory *P.M. Cohn*
Introductory Mathematics: Algebra and Analysis *G. Smith*
Linear Functional Analysis *B.P. Rynne and M.A. Youngson*
Matrix Groups: An Introduction to Lie Group Theory *A. Baker*
Measure, Integral and Probability *M. Capińksi and E. Kopp*
Multivariate Calculus and Geometry *S. Dineen*
Numerical Methods for Partial Differential Equations *G. Evans, J. Blackledge, P. Yardley*
Probability Models *J. Haigh*
Real Analysis *J.M. Howie*
Sets, Logic and Categories *P. Cameron*
Symmetries *D.L. Johnson*
Topics in Group Theory *G. Smith and O. Tabachnikova*
Topologies and Uniformities *I.M. James*
Vector Calculus *P.C. Matthews*

N.M.J. Woodhouse

Special Relativity

With 17 Figures

Springer

Nicholas M.J. Woodhouse, MA, MSc, PhD
Mathematical Institute, Oxford University, 24-29 St Giles', Oxford OX1 3LB, UK

Cover illustration elements reproduced by kind permission of:
Aptech Systems, Inc., Publishers of the GAUSS Mathematical and Statistical System, 23804 S.E. Kent-Kangley Road, Maple Valley, WA 98038, USA. Tel: (206) 432 - 7855 Fax (206) 432 - 7832 email: info@aptech.com URL: www.aptech.com
American Statistical Association: Chance Vol 8 No 1, 1995 article by KS and KW Heiner 'Tree Rings of the Northern Shawangunks' page 32 fig 2
Springer-Verlag: Mathematica in Education and Research Vol 4 Issue 3 1995 article by Roman E Maeder, Beatrice Amrhein and Oliver Gloor 'Illustrated Mathematics: Visualization of Mathematical Objects' page 9 fig 11, originally published as a CD ROM 'Illustrated Mathematics' by TELOS: ISBN 0-387-14222-3, German edition by Birkhauser: ISBN 3-7643-5100-4.
Mathematica in Education and Research Vol 4 Issue 3 1995 article by Richard J Gaylord and Kazume Nishidate 'Traffic Engineering with Cellular Automata' page 35 fig 2. Mathematica in Education and Research Vol 5 Issue 2 1996 article by Michael Trott 'The Implicitization of a Trefoil Knot' page 14.
Mathematica in Education and Research Vol 5 Issue 2 1996 article by Lee de Cola 'Coins, Trees, Bars and Bells: Simulation of the Binomial Process' page 19 fig 3. Mathematica in Education and Research Vol 5 Issue 2 1996 article by Richard Gaylord and Kazume Nishidate 'Contagious Spreading' page 33 fig 1. Mathematica in Education and Research Vol 5 Issue 2 1996 article by Joe Buhler and Stan Wagon 'Secrets of the Madelung Constant' page 50 fig 1.

British Library Cataloguing in Publication Data
Woodhouse, Nicholas, 1949-
Special relativity. - (Springer undergraduate mathematics series)
1. Special relativity (Physics)
I. Title
530.1'1
ISBN 1852334266

Library of Congress Cataloging-in-Publication Data
Woodhouse, N.M.J. (Nicholas Michael John), 1949-
Special relativity / N.M.J. Woodhouse.
p. cm. -- (Springer undergraduate mathematics series, ISSN 1615-2085)
Includes bibliographical references and index.
ISBN 1-85233-426-6 (acid-free paper)
1. Special relativity (Physics) I. Title. II. Series.
QC173.65.W66 2002
530.11—dc21 2002190890

Springer Undergraduate Mathematics Series ISSN 1615-2085
ISBN 1-85233-426-6 Springer-Verlag London Berlin Heidelberg
a member of BertelsmannSpringer Science+Business Media GmbH
http://www.springer.co.uk

Typesetting: Camera ready by the author
Printed and bound at the Athenæum Press Ltd., Gateshead, Tyne & Wear
12/3830-543210 Printed on acid-free paper SPIN 10791491

Preface

Mathematics courses in British universities used to contain a very large measure of physical applied mathematics. Undergraduates would come to relativity and quantum theory in their final years with a strong background in mechanics, fluid dynamics, Newtonian gravitation, and classical electromagnetism. That, in turn, would have been built on a mathematical approach to physics in schools. Recent years, however, have seen a widening in the range of undergraduate applied mathematics, in statistics, computation, mathematical biology, finance, and modelling. There are new opportunities, but there is also a new challenge: how to teach mathematical physics in a way that does not require either early and undue specialization or else an endpoint that falls well short of the revolution at the beginning of the last century.

The physical applications must retain a place in the core of mathematics education. Not only do they provide a paradigm for all mathematical models, but they are also central to the historical development of ideas and a key to understanding an important aspect of the truth of mathematics. Without them, one loses sight of the roots of the subject and of what Wigner called 'its unreasonable effectiveness' [13]. But a course that does not go beyond nineteenth-century physics is unlikely to fire the imagination; and one that introduces the modern theories without the necessary background runs the risk of trivializing them, of presenting deep ideas as a sequence of neatly packaged facts to be reproduced in examination, and of missing the central point that mathematics frees our minds to explore the world in which we live beyond the limits of our physical intuition. To understand relativity, in particular, one must develop a clear mental picture of space–time, and not simply a facility for implementing Lorentz transformations. What a mathematician brings to this task is the ability to think in a coherent way about abstract structures.

This book attempts to present the special theory of relativity in a way that draws on the mathematical experience of undergraduates, but requires neither an extensive background in classical mathematical physics nor a long prologue devoted to the development of new tools, such as tensor analysis. That is not to say that the theory is presented in the style of pure mathematics, nor that a more extensive background is not helpful; simply that it is written for students who are more familiar with abstract linear algebra than with electromagnetism. In one sense, this approach makes the task easier: if one comes across Maxwell's equations for the first time in the context of the relativity problem, then Einstein's resolution appears all the more natural.

In writing it, I have been heavily influenced by those from whom I first learnt relativity, either directly or through their writing; particularly by Roger Penrose, Felix Pirani, and John Wheeler, and by the books by W. Rindler [8], [9] and J. L. Synge [10]. It is not always easy to track down and acknowledge explicitly the source of ideas in a subject such as this, and I am conscious, in particular, that many of the problems included here have been 'borrowed' over the years from the work of others. Some, I know, are based on problems from Rindler's book [8], and these are marked with a dagger (†). Others (marked $^+$) are adapted from Oxford examination papers; their setters are cloaked with anonymity. Chapters and sections marked with an asterisk (*) touch on harder ideas, and can be omitted. The quotation from Galilieo on pp. 2–3 is reprinted by kind permission of the University of California Press. (Copyright © 1952, 1962, 1967 Regents of the University of California.)

I am grateful to many colleagues and students in Oxford for pointing out errors in various versions of lecture notes from which the book has developed.

Oxford, April 2002 NMJW

Contents

1
Relativity in Classical Mechanics

1.1 Frames of Reference

In order to describe the motion of a system of particles in classical mechanics, it is necessary first to choose a frame of reference—that is, an origin and a set of right-handed Cartesian axes. Sometimes there is a natural choice. In a projectile problem, for example, it is sensible to take the z-axis to point directly upwards and to pick out the origin and the direction of the x-axis from the initial conditions. Such choices become embedded in our physical intuition to the extent that it is difficult to think clearly about a problem in an unconventional frame—with the z-axis pointing downwards, for example, or at an angle to the horizontal.

But the difficulties are not fundamental. They come from the overlaying of physical law with conventions of description. The laws of classical mechanics, Newton's three laws of motion and the inverse-square law of gravity, are true whatever choice we make. The differential equations of motion derived from them may look more complicated if the natural choices are ignored, and it may be harder to see how to solve them; but they still determine the behaviour of the system.

This observation reflects two very important properties of space. It is *homogeneous* and *isotropic*. In other words, there is no preferred place (homogeneity) and no preferred direction (isotropy). The laws of mechanics take the same form wherever you are and in whatever direction you look. There is no reason beyond convention and convenience to make one choice rather than another. To put the conclusion positively, the laws of classical mechanics are invariant under translation of the origin and under rotation of the axes—space has rotational and translational symmetry.

1.2 Relativity

The principle of relativity emerges from thinking about another freedom in the choice of frame—its motion. We are used to taking the earth as defining a standard of rest and to building into our intuition about the behaviour of mechanical systems the notion that the frame of reference is *fixed*; that is that it is fixed relative to the earth. This is reasonable in the context of terrestrial problems; but it has to be set aside, sometimes in a way that is quite hard to picture, in problems that involve motion over large distances. The rotation of the earth, for example, can be discounted in laboratory experiments, although its effects can just be observed in the behaviour of Foucault's pendulum. It is not ignorable in the large-scale behaviour of the weather (it causes air to circulate in an anti-clockwise direction around a region of low pressure in the Northern hemisphere); and it would be an unnecessary handicap to insist on working in a frame fixed to the earth in predicting the trajectory of a comet or planetary probe.

The idea that we are also free to choose the motion of the frame predates the formulation of the laws of motion. It is put very clearly in Galileo's *Dialogue concerning the two chief world systems* [3].

> Shut yourself up with some friend in the main cabin below the decks on some large ship, and have with you there some flies, butterflies, and other small flying animals. Have a large bowl of water with some fish in it; hang up a bottle that empties drop by drop into a wide vessel beneath it. With the ship standing still, observe carefully how the little animals fly with equal speed to all sides of the cabin. The fish swim indifferently in all directions; the drops fall into the vessel beneath; and, in throwing something to your friend, you need throw it no more strongly in one direction than another, the distances being equal; jumping with your feet together, you pass equal spaces in every direction. When you have observed all these things carefully (though there is no doubt that when the ship is standing still everything must happen in this way), have the ship proceed with any speed you like, so long as the motion is uniform and not fluctuating this way and that. You will discover not the least change in all the effects named, nor could you tell from any of them whether the ship was moving or

> standing still. In jumping, you will pass on the floor the same spaces as before, nor will you make larger jumps towards the stern than towards the prow even though the ship is moving quite rapidly, despite the fact that during the time that you are in the air the floor under you will be going in a direction opposite to your jump. In throwing something to your companion, you will need no more force to get it to him whether he is in the direction of the bow or the stern, with yourself situated opposite. The droplets will fall as before into the vessel beneath without dropping toward the stern, although while the drops are in the air, the ship runs many spans. The fish in their water will swim toward the front of their bowl with no more effort than toward the back ... the flies and butterflies will continue their flights indifferently toward every side....

Thus Galileo is observing (ahead of their precise statement by Newton) that the laws of motion are the same in a 'fixed' frame and in a frame moving with the ship, provided that the motion is uniform. The proviso is important. It is not true, and it is not asserted, that the acceleration or rotation of the frame is unimportant; simply that its uniform motion in a straight line cannot be detected within the ship.

We now have more modern illustrations of the Galileo's point. A ball thrown across a railway carriage running on a smooth track, or across the cabin of an aeroplane in steady flight, will behave in the same way as in a fixed laboratory. The simplest way to describe its motion is to choose a frame of reference moving with the train or plane, rather than one fixed relative to the earth.

The idea comes into sharper focus if we remove the earth altogether from the picture. Imagine two spaceships passing each other in empty space. In each, the passengers think that they are at rest and that it is the other spaceship that is moving. Is there some physical test that will determine who is right? Certainly there is if one is accelerating and the other is not since it is possible to *feel* acceleration. But if both are moving uniformly at constant speed, then there is not; no device within one of the spaceships will detect uniform motion, only motion relative to some external standard, such as the earth or the 'fixed stars'. The *principle of relativity* turns this from a negative to a positive statement.

Principle of Relativity

In classical mechanics, all frames of reference in uniform motion are equivalent.

As with the origin and the orientation of the axes, the freedom to choose the motion of the frame can be expressed as a symmetry, not in this case of space, but of *space-time*—space and time considered together.

To understand how this works, and to introduce some basic notation, we must first make the preceding discussion a more little precise by looking in a more formal way at Newton's laws and their invariance.

1.3 Frames of Reference

Let us choose a frame R—that is, an origin together with a (right-handed) set of Cartesian axes. So as not to prejudge the issue, we must avoid the temptation to think of R as 'fixed in space'.

The *position*, *velocity*, and *acceleration* of a particle relative to R are the vectors $\boldsymbol{r}$, $\boldsymbol{v}$, and $\boldsymbol{a}$ with components

$$(x, y, z), \quad (\dot{x}, \dot{y}, \dot{z}), \quad (\ddot{x}, \ddot{y}, \ddot{z}), \tag{1.1}$$

where x, y, z are the coordinates of the particle and dot is the derivative with respect to time.

If R' is a second frame, which may be moving or rotating relative to R, then the coordinates of a particle in the two frames are related by an affine linear transformation[1]

$$\begin{pmatrix} x \\ y \\ z \end{pmatrix} = H \begin{pmatrix} x' \\ y' \\ z' \end{pmatrix} + T \tag{1.2}$$

where

$$H = \begin{pmatrix} H_{ij} \end{pmatrix} = \begin{pmatrix} H_{11} & H_{12} & H_{13} \\ H_{21} & H_{22} & H_{23} \\ H_{31} & H_{32} & H_{33} \end{pmatrix} \tag{1.3}$$

is a proper orthogonal matrix, representing a rotation and

$$T = \begin{pmatrix} T_i \end{pmatrix} = \begin{pmatrix} T_1 \\ T_2 \\ T_3 \end{pmatrix} \tag{1.4}$$

is a column vector, representing a translation. The H_{ij}s and T_is are functions of time t.

By differentiating twice, we find that the components $(\ddot{x}, \ddot{y}, \ddot{z})$ of the acceleration $\boldsymbol{a}$ of the particle relative to R are related to the components $(\ddot{x}', \ddot{y}', \ddot{z}')$ of the acceleration $\boldsymbol{a}'$ relative to R' by

$$\begin{pmatrix} \ddot{x} \\ \ddot{y} \\ \ddot{z} \end{pmatrix} = H \begin{pmatrix} \ddot{x}' \\ \ddot{y}' \\ \ddot{z}' \end{pmatrix} + 2\dot{H} \begin{pmatrix} \dot{x}' \\ \dot{y}' \\ \dot{z}' \end{pmatrix} + \ddot{H} \begin{pmatrix} x' \\ y' \\ z' \end{pmatrix} + \ddot{T}, \tag{1.5}$$

[1] An *affine transformation* $X \mapsto AX + B$ is a combination of a linear transformation and a translation. Here X is a column vector, A is nonsingular square matrix, and B is a fixed column vector. The map $X \mapsto AX$ is the *associated linear transformation.*

where $\dot{H} = (\dot{H}_{ij})$, $\dot{T} = (\dot{T}_i)$, and so on. Thus $\boldsymbol{a} = \boldsymbol{a}'$ for every possible motion of the particle if and only if $\dot{H} = 0$ and $\ddot{T} = 0$. That is, if and only if R' is moving relative to R without rotation or acceleration.

There is an important lesson to note here in passing. When there is more than one frame in play, one must keep clear the distinction between a vector and its components. A *vector*, denoted by a bold letter, is a quantity with magnitude and direction. It is a geometric object, pictured as an arrow in space. The basic example is the displacement from a point A to the point B, which we picture as an arrow drawn from A to B. Vectors can be added (by the parallelogram rule) and multiplied by scalars (real numbers) without any need to mention a frame of reference or components. But if we *are* given a frame of reference then a vector $\boldsymbol{X}$ can be represented by its three components a, b, c. If $\boldsymbol{X}$ is the displacement from A to B, then you get to B from A by moving a units in the x-direction, b units in the y-direction, and c units in the z-direction. Addition is given by adding components.

When the frame is changed by rotating the axes and moving the origin, as in (1.2), the components of $\boldsymbol{X}$ undergo the linear transformation,

$$\begin{pmatrix} a \\ b \\ c \end{pmatrix} \mapsto \begin{pmatrix} a' \\ b' \\ c' \end{pmatrix}, \quad \text{where} \quad \begin{pmatrix} a \\ b \\ c \end{pmatrix} = H \begin{pmatrix} a' \\ b' \\ c' \end{pmatrix} \tag{1.6}$$

while $\boldsymbol{X}$ itself remains unchanged. The 3×1 matrix with entries a, b, c is a *column vector*, but not a vector in the geometric sense; one can run into trouble by equating $\boldsymbol{X}$ to the column vector of its components, although this is frequently done, and is a convenient abuse of notation in contexts in which there is just one frame in view.

1.4 Newton's Laws

Newton's three laws of motion can be stated as follows.

- In the absence of forces, a body moves in a straight line at constant speed.
- The acceleration of a body of mass m is related to the force acting on it by $\boldsymbol{F} = m\boldsymbol{a}$.
- To every action there is an equal and opposite reaction.

Their familiarity and simplicity conceal subtle questions. What is 'mass' and what is 'force'? How can force be determined, other than through the acceleration it produces on a standard mass? What is the content of the second law beyond a definition of force?

We shall return to the definition of mass in Chapter 7. For the moment we shall concentrate on the first law, since it does not involve mass or force, except in the self-explanatory negative sense that force should be absent; a body without anything pushing or pulling it will move in a straight line at constant speed. We shall also concentrate on its application to point particles, so that we do not have to worry about the internal structure of extended bodies. (Even taking the meaning of 'absence of force' as self-evident raises problems in the presence of gravity, since gravity cannot be turned off—it affects all matter equally in proportion to its mass. Thus in a gravitational field, there is no obvious experiment that one could do to check whether or not the first law is true. It is through thinking about this issue that Einstein was led to formulate the *general theory of relativity*.)

Newton's first law is a statement about the existence of a class of frames of reference as well about the behaviour of particles. A particle that has zero acceleration relative to one frame of reference will have nonzero acceleration in a second frame that is accelerating or rotating relative to the first. On the other hand, if the first law holds in one frame of reference, then it holds in every frame of reference related to it by a transformation (1.2) in which

$$\dot{H} = 0, \qquad \ddot{T} = 0\,.$$

We can thus restate the first law as follows.

Law of Inertia

There exists a class of frames of reference relative to which the motion of a particle not subject to any force is in a straight line at constant speed.

If R is such a frame, then so is any other frame R' which is related to R by a transformation (1.2) in which H and $\dot{T}$ are constant.

Definition 1.1

An *inertial frame* is a frame of reference R in which the law of inertia holds.

The Galilean *principle of relativity* is that, as far as the laws of mechanics are concerned, *all inertial frames are on an equal footing*. Newton's laws hold in all inertial frames and no mechanical phenomenon will allow any one frame of reference to be distinguished as 'at rest'.

Aside

A point that can cause confusion in reconciling this principle with ideas from elementary mechanics is that a frame fixed on earth is *not* inertial, but only approximately so; the approximation is to ignore the earth's rotation. In problems where the rotation matters, for example in the analysis of Foucault's pendulum, it is helpful to treat a terrestrial frame as if it were inertial, but to introduce correction terms, the 'fictitious' centrifugal and Coriolis forces, to take account of the rotation. But these are not real forces, and the frame is not really inertial.

Example 1.2 (Slingshot Effect)

This is an example that illustrates the power of the principle in solving dynamical problems. Suppose that a spacecraft of small mass M in free-fall has velocity $\boldsymbol{v}$ when it is at a large distance from a stationary planet of very much larger mass m. Provided that it does not hit the planet, it will pass the planet on a hyperbolic orbit and escape again to infinity. If it has velocity $\boldsymbol{v}'$ when it is again at a large distance from the planet, then by conservation of energy in the frame of the planet, we shall have $|\boldsymbol{v}| = |\boldsymbol{v}'|$. However, the direction of motion will have changed. It is not hard to show by using the standard theory of orbits under an inverse-square law force that

$$\boldsymbol{v}.\boldsymbol{v}' = -v^2 \cos 2\alpha \,,$$

where $v = |\boldsymbol{v}|$, and that, if d is the distance of closest approach to the centre of the planet, then

$$\cos\alpha = \frac{Gmd}{Gmd + v^2 d^2} \,.$$

If d is small, then $\alpha \sim 0$, and the trajectory of the spacecraft is reversed; if d is large, then $\alpha \sim \pi/2$ and the deflection is small.

What happens when the spacecraft passes close to a moving planet? To find out, we first consider the motion in a frame R in which the planet is at rest and then transform the results to a frame R' in which the planet is moving. Denote by $\boldsymbol{v}$ and $\boldsymbol{v}'$ the respective velocities relative to R of the spacecraft before and after the encounter; and let $\boldsymbol{u}$ denote the velocity of the planet relative to the frame R'. Then the craft has velocity $\boldsymbol{v} + \boldsymbol{u}$ relative to R' before the encounter and velocity $\boldsymbol{v}' + \boldsymbol{u}$ after the encounter. From conservation of energy in the frame R, we have

$$\tfrac{1}{2}M\boldsymbol{v}'.\boldsymbol{v}' = \tfrac{1}{2}M\boldsymbol{v}.\boldsymbol{v} \,.$$

So relative to R', the spacecraft gains kinetic energy

$$E = \tfrac{1}{2}M(\boldsymbol{v}' + \boldsymbol{u}).(\boldsymbol{v}' + \boldsymbol{u}) - \tfrac{1}{2}M(\boldsymbol{v} + \boldsymbol{u}).(\boldsymbol{v} + \boldsymbol{u}) = M\boldsymbol{u}.(\boldsymbol{v}' - \boldsymbol{v}) .$$

In the extreme case of small d, we have $\boldsymbol{v}' \sim -\boldsymbol{v}$ and that E is maximal when $\boldsymbol{u}$ is in the opposite direction to $\boldsymbol{v}$. Of course the gain is matched by a loss in energy of the planet, although we have neglected the change in the velocity of the planet by making the assumption that m is very much greater than M.

This *slingshot effect* is used to boost the energy of space probes travelling to the outer planets by encounters on the way with Jupiter or Saturn.

Newton's Fourth Law

In the manuscript *De motu corporum in mediis regulariter cedentibus* that Newton wrote two and half years before the laws of motion appeared in his *Philosophiae naturalis principia mathematica*, he had not three, but six laws of motion. The fourth was the principle of relativity.

> The relative motion of bodies in a given space is the same whether the space is absolutely at rest or moves in a straight line without rotation.

Newton realized that the fourth law was a consequence of the first three (the three laws we know today); but he had other reasons for believing in an absolute standard of rest, which remains 'always similar and immovable'. In the *Principia*, the fourth law is reduced to the status of a corollary to the laws of motion.

1.5 Galilean Transformations

If we have $\dot{H} = 0$, $\ddot{T} = 0$ in (1.2), then Newton's laws hold in R' if and only if they hold in R. We can also combine this transformation with a change in the origin of the time coordinate without affecting the validity of the laws of motion. The result is a *Galilean transformation* that changes the space and time coordinates from t, x, y, z to t', x', y', z', where

$$t = t' + \text{constant} ,$$

and where x, y, z and x', y', z' are related by (1.2), with H a constant matrix, and

$$T = vt' + c \tag{1.7}$$

for some constant column vectors v and c. In four-dimensional form, the definition becomes the following.

Definition 1.3

A *Galilean transformation* is a coordinate transformation

$$\begin{pmatrix} t \\ x \\ y \\ z \end{pmatrix} = \begin{pmatrix} 1 & 0 & 0 & 0 \\ v_1 & H_{11} & H_{12} & H_{13} \\ v_2 & H_{21} & H_{22} & H_{23} \\ v_3 & H_{31} & H_{32} & H_{33} \end{pmatrix} \begin{pmatrix} t' \\ x' \\ y' \\ z' \end{pmatrix} + C \tag{1.8}$$

where v, H, and C are constant; $v = (v_i)$ is a column vector of length 3, $H = (H_{ij})$ is a 3×3 proper orthogonal matrix, and C is a column vector of length 4.

Galilean transformations preserve Newton's laws in the sense that if two coordinate systems are related by a Galilean transformation, then the laws hold in one system if and only if they hold in the other.

Rotations

If $C = 0$, $v = 0$, then

$$\begin{pmatrix} t \\ x \\ y \\ z \end{pmatrix} = \begin{pmatrix} 1 & 0 & 0 & 0 \\ 0 & H_{11} & H_{12} & H_{13} \\ 0 & H_{21} & H_{22} & H_{23} \\ 0 & H_{31} & H_{32} & H_{33} \end{pmatrix} \begin{pmatrix} t' \\ x' \\ y' \\ z' \end{pmatrix} .$$

In this case, $t = t'$ and the x, y, z and x', y', z' coordinates are related by a rotation of the axes. The frames are at rest relative to each other.

Boosts

If $H = 1$, $C = 0$, then $t = t'$ and

$$\begin{pmatrix} x \\ y \\ z \end{pmatrix} = \begin{pmatrix} x' \\ y' \\ z' \end{pmatrix} + \begin{pmatrix} v_1 t \\ v_2 t \\ v_3 t \end{pmatrix} .$$

The axes are parallel and the origins coincide at $t = 0$. The frame R' moves with constant velocity (v_1, v_2, v_3) relative to R.

Translations

If $v = 0$, $H = 1$, then the coordinate systems are related by a translation of the origin and a resetting of the zero of t.

Every Galilean transformation is a combination of a rotation, a boost, and a translation. In the classical picture, the coordinate systems of any two inertial frames are related by a Galilean transformation, and any frame related to an inertial frame by a Galilean transformation is itself inertial. There is nothing in the laws of classical mechanics that picks out a particular inertial frame. The principle of relativity is that all inertial frames are equivalent.

1.6 Mass, Energy, and Momentum

In Newtonian mechanics, mass and force are invariant—they are the same in every inertial frame. Acceleration is unchanged by Galilean transformations, and so Newton's second law $\boldsymbol{F} = m\boldsymbol{a}$ holds in all frames if it holds in one. This is consistent with an intuitive idea of mass as 'quantity of matter'. Mass is conserved. It is neither created nor destroyed in mechanical processes, and its value is unchanged by motion.

Energy and momentum, on the other hand, are not invariant. A particle at rest has no kinetic energy or momentum; but when it is observed from a moving frame, it acquires both. In the slingshot (Example 1.2), the spacecraft's kinetic energy was unchanged in the rest frame of the planet,[2] but not in the frame in which the planet had velocity $\boldsymbol{u}$. At first sight, this appears to violate conservation of energy: it seems that energy has been conserved in one frame (the rest frame of the planet) but not in the other. This is only because the working in the example took no account of the change in the kinetic energy of the planet itself—a reasonable approximation to make if the planet has much larger mass than the spacecraft. In both frames, the sum of the initial kinetic energies of the planet and the spacecraft is the same as the sum of final kinetic energies of the planet and spacecraft; but their individual kinetic energies in the two frames are different.

A simple type of problem in which to see how this works is one involving only *collisions*. We shall use this word in a rather general sense to mean 'interactions between particles at a point'. It includes collisions between particles in the conventional sense, as well as coalescence (two or more particles hitting each other and sticking together) and fragmentation (a particle breaking up into two

[2] The *rest frame* of a moving body is the frame in which it is at rest.

or more new particles). Thus the number of 'incoming' particles in a collision need not be the same as the number of 'outgoing particles'. Between collisions, the particles move in straight lines at constant speed.

In Newtonian mechanics, the behaviour of particles in a collision is governed by three laws.

Conservation of Mass

If the masses of the incoming particles are m_1, m_2, ..., m_k and those of the outgoing particles are $m_{k+1}, m_{k+2}, \dots, m_n$, then

$$\sum_{1}^{k} m_i = \sum_{k+1}^{n} m_i \,,$$

although the number k of incoming particles need not be the same as the number $n - k$ of outgoing particles because of fragmentation and coalescence.

Conservation of Momentum

If the velocities of the incoming particles are $\boldsymbol{v}_1, \boldsymbol{v}_2, \dots, \boldsymbol{v}_k$ and those of the outgoing particles are $\boldsymbol{v}_{k+1}, \boldsymbol{v}_{k+2}, \dots, \boldsymbol{v}_n$, then

$$\sum_{1}^{k} m_i \boldsymbol{v}_i = \sum_{k+1}^{n} m_i \boldsymbol{v}_i \,.$$

Conservation of Energy

This is less straightforward since 'collisions' can involve the interchange of kinetic and internal energy. Energy is also conserved, but kinetic energy need not be since there is generally an interchange between different forms of energy—unless the collision is elastic. In an inelastic collision, kinetic energy is transformed into heat; in an explosion, chemical energy is released as the kinetic energy of the fragments. If the total internal energy of the incoming particles is I_{in} and that of the outgoing particles is I_{out}, then

$$I_{\text{in}} + \sum_{1}^{k} \tfrac{1}{2} m_i \boldsymbol{v}_i . \boldsymbol{v}_i = I_{\text{out}} + \sum_{k+1}^{n} \tfrac{1}{2} m_i \boldsymbol{v}_i . \boldsymbol{v}_i \,.$$

When we observe the same collision from a different inertial frame of reference, the velocities of the particles become $\boldsymbol{v}_i + \boldsymbol{u}$, where $\boldsymbol{u}$ is the velocity of the first frame relative to the second. In the second frame, the momentum and kinetic energy of the ith particle are, respectively,

$$m_i(\boldsymbol{v}_i + \boldsymbol{u}) = m_i\boldsymbol{v}_i + m_i\boldsymbol{u}$$

and

$$\tfrac{1}{2}m_i(\boldsymbol{v}_i + \boldsymbol{u}).(\boldsymbol{v}_i + \boldsymbol{u}) = \tfrac{1}{2}m_i\boldsymbol{v}_i.\boldsymbol{v}_i + m_i\boldsymbol{v}_i.\boldsymbol{u} + \tfrac{1}{2}m_i\boldsymbol{u}.\boldsymbol{u}\,.$$

Thus if M is the total mass, $\boldsymbol{P}$ is the total momentum in the first frame, and K is the total kinetic energy before the collision in the first frame, then in the second frame the total momentum and total energy are, respectively,

$$\boldsymbol{P} + M\boldsymbol{u} \qquad \text{and} \qquad I_{\text{in}} + K + \boldsymbol{P}.\boldsymbol{u} + \tfrac{1}{2}M\boldsymbol{u}.\boldsymbol{u}\,.$$

So conservation of momentum and energy in the first frame, together with conservation of mass, imply conservation of momentum and conservation of energy in the second frame. It is also true that if energy is conserved for every $\boldsymbol{u}$, then $\boldsymbol{P}$ and M must also be conserved. So *conservation of energy in every inertial frame implies conservation of momentum and conservation of mass*, provided that we assume that the internal energy of a particle is the same in whatever frame it is measured.

1.7 Space–time

To summarize: in classical mechanics, we have the following basic principles.

- There is a preferred class of frames of reference—the *inertial frames.*
- A frame is inertial if and only if Newton's first law holds for every particle not influenced by any forces.
- An inertial frame, together with a choice of zero for t, determines an *inertial coordinate system* t, x, y, z.
- The inertial coordinate systems of two inertial frames are related by a Galilean transformation.
- Newton's three laws hold for any mechanical system in an inertial frame.
- Mass is independent of motion, and is conserved.

We shall see that the first three statements also hold in Einstein's theory of relativity; but the fourth is only approximately true when the two frames are moving relative to each other at a speed much less than c, the speed of light. Compatibility with Maxwell's equations implies that the Galilean transformation must be replaced by the *Lorentz transformation*. This is the same as the Galilean transformation in the limit of slow relative motion, but has rather different properties when the two frames are moving relative to each other at speeds comparable with c. The final statement will also require modification.

Some of the disorientation that results from replacing the classical picture by Einstein's is unavoidable; Einstein's theory is different, and implies that space and time have properties that run counter to intuition. But some is artificial, and is generated by the requirement to think about problems of space and time in an unfamiliar way: it is not just that Einstein's relativity theory is unfamiliar—the notion of relativity is itself something that we do not often think about because the standard of rest determined by the earth is so much part of our physical intuition.

It is helpful, therefore, to think about the implications of Galileo's principle in terms of ideas that we shall meet later in Einstein's theory. A central one is the notion of a *space-time diagram*. In geometry, it is often helpful to form mental images of objects in two- or three-dimensional space without making a particular choice of Cartesian axes. One thinks of the axes as being added to the picture rather than intrinsic to it; and one is free to make the addition in whichever way is most convenient to solve the problem at hand. In the same way, it is helpful in thinking about relativity (in a classical context as well as in Einstein's theory) to form a mental image of space–time as a four-dimensional space, to which the space and time axes, and the choice of coordinates, are added as convenient. Different choices of axes correspond to different choices of inertial frame.

A 'point' of this four-dimensional space is an *event*—a particular location at a particular time. Events are labelled by t (time) and by the three Cartesian coordinates x, y, z, giving a system of four coordinates t, x, y, z on space–time.

Definition 1.4

Space-time is the set of all events. An *event* is a particular point at a particular time.

Let A, B be two events in space–time and consider the following statements.

- A and B are simultaneous;

- B happens time t after A;

– A and B are simultaneous and separated by a distance D;

– A and B happen in the same place (at different times);

– A and B are separated by distance D, but happen at different times.

The first three of these are *invariant*—they do not depend on the particular choice of frame. If one of them is true in one system of inertial coordinates, then it is true in every system. The last two, however, are not invariant. Suppose, for example, that A is '2 pm in Oxford' and B is '3 pm in Oxford'. Then in a frame in which the earth is fixed, A and B happen in the same place; but in a frame in which the sun is at rest, A and B are separated by some 70,000 miles, since the earth is moving relative to the sun at some 19 miles per second.

This is an aspect of the classical picture which sometimes challenges intuition: the concept of *location* is relative. The statement that 'such and such an event happened here' requires a choice of frame of reference. It may be true with one choice, but false with another. Our intuition fails us here because it equates the idea of 'place' with 'place on earth': we build into our everyday thinking a particular frame of reference fixed relative to the earth, and forget that in a moving frame, two events at the same location relative to the earth happen in different 'places'. Similarly, we do not pause to consider that the idea of 'distance' between non-simultaneous events is also relative—it depends on a choice of frame.

With one space dimension suppressed, we can picture space–time as in Figure 1.1, which is an example of a *space–time diagram*. Here the time axis points upwards, and the horizontal planes represent space at different times. The straight line L represents the history of a particle moving in a straight line at constant speed (the greater the slope, the *lower* the speed). A curve in

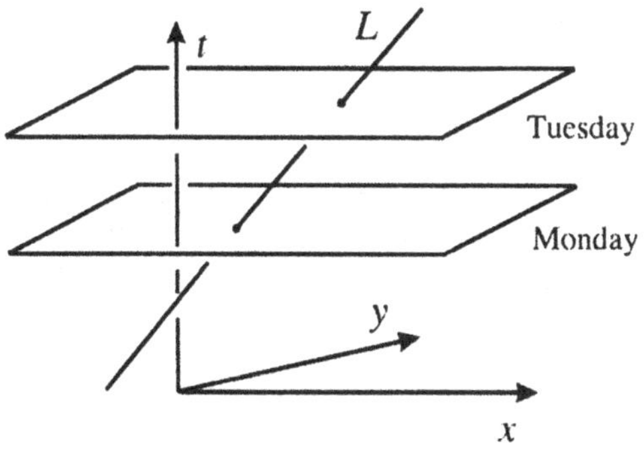

Figure 1.1 The space–time of Galilean relativity

space–time representing the history of a particle in general motion is called a *worldline.*

With only one space dimension, a Galilean transformation takes the form

$$\begin{pmatrix} t \\ x \end{pmatrix} = \begin{pmatrix} 1 & 0 \\ v & 1 \end{pmatrix} \begin{pmatrix} t' \\ x' \end{pmatrix} + \begin{pmatrix} c_0 \\ c_1 \end{pmatrix} .$$

This is illustrated in Figure 1.2. Note that the lines of constant t and the lines

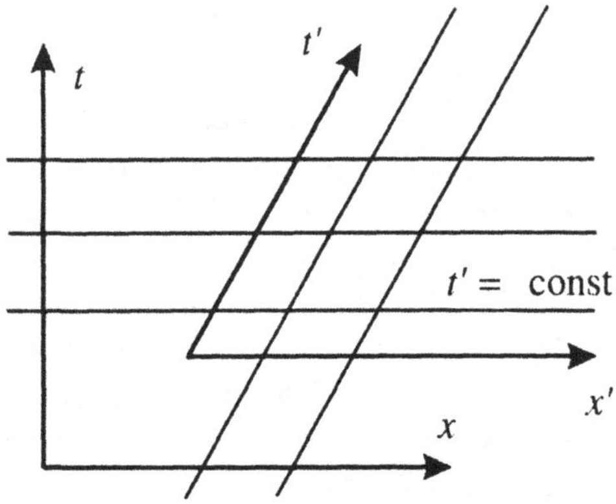

Figure 1.2 A Galilean transformation in two-dimensional space–time

of constant t' coincide, reflecting the invariance of simultaneity. The central difference between the space–time of classical physics and Einstein's space–time is that simultaneity is *not* invariant in Einstein's theory: it is also relative.

Space–time diagrams are sometimes useful in other contexts—for example, for scheduling trains on a single-track railway line. The following problem provides a nice illustration.

Problem 1.5

Four ghosts travel in straight lines at different constant speeds across a flat field. Five of the six pairs pass through each other at different times. Show that the sixth pair also pass through each other.

Solution

We have a three-dimensional space–time with coordinates t, x, y, where x, y are Cartesian coordinates on the field, and t is time. In this, the worldlines of the ghosts are straight lines (since they travel in straight lines at constant speeds).

Three of the ghosts all pass through each other, so their worldlines intersect in three distinct points (they are distinct because the ghosts pass through each other at different times). These three lines must therefore be coplanar. The worldline of the fourth ghost intersects two of the other lines, so it must also lie in the plane. It must therefore intersect all three. So all the ghosts pass through each other. □

1.8 *Galilean Symmetries

A rotation or translation in space can be thought of in two ways: as a transformation between Cartesian coordinate systems (a *passive* transformation in which we leave space alone, but change the way in which its points are labelled by coordinates) or as a mapping of space to itself (an *active* transformation in which the points themselves move).

Definition 1.6

Let $\mathbb{E}$ denote three-dimensional Euclidean space. An *isometry* is a map $\iota : \mathbb{E} \to \mathbb{E}$ which is given in Cartesian coordinates by

$$\begin{pmatrix} x \\ y \\ z \end{pmatrix} \mapsto H \begin{pmatrix} x \\ y \\ z \end{pmatrix} + T,$$

where T is a constant column vector and H is a 3×3 orthogonal matrix.

If a mapping takes this form for one choice of Cartesian coordinates, then it does so for any other choice (but with different H and T). It can be shown that a map $\iota : \mathbb{E} \to \mathbb{E}$ is an isometry if and only if it preserves distances between pairs of points. If we think of 'distance between points' as part of the structure of $\mathbb{E}$, then the isometries are the maps that preserve this structure—they are the *symmetries* of space.

An isometry can be *proper* or *improper*, according to whether $\det H = 1$ or $\det H = -1$. The proper isometries preserve the *orientation* of space: they map right-handed triads of vectors to right-handed triads, while improper isometries reverse orientation.

We can similarly think of Galilean transformations either as passive transformations between coordinate systems or as active transformations of space–time.

Definition 1.7

Let $\mathbb{G}$ denote the four-dimensional space of events. An active *Galilean symmetry* is a map $\gamma : \mathbb{G} \to \mathbb{G}$ given in inertial coordinates by

$$\begin{pmatrix} t \\ x \\ y \\ z \end{pmatrix} \mapsto \begin{pmatrix} 1 & 0 & 0 & 0 \\ v_1 & H_{11} & H_{12} & H_{13} \\ v_2 & H_{21} & H_{22} & H_{23} \\ v_3 & H_{31} & H_{32} & H_{33} \end{pmatrix} \begin{pmatrix} t \\ x \\ y \\ z \end{pmatrix} + C$$

for some constant orthogonal matrix H, constant v_1, v_2, v_3, and constant column vector C.

If t_1, x_1, y_1, z_1 and t_2, x_2, y_2, z_2 are the coordinates of a pair of events, then a Galilean symmetry preserves (i) the time interval $t_2 - t_1$ between any pair of events and (ii) the distance

$$\sqrt{(x_2 - x_1)^2 + (y_2 - y_1)^2 + (z_2 - z_1)^2}$$

provided that the events are *simultaneous*—that is, provided that interval $t_2 - t_1$ vanishes. It also preserves the orientation of space if $\det H = 1$. The Galilean symmetries are the symmetries of classical space–time.

It is not true, however, that Galilean symmetries are characterized by the fact that they preserve time intervals and distance for simultaneous events; any transformation of the same form as that in the definition, but in which H depends on t in an arbitrary way, will have this property. But it is true that any *affine* transformation $\gamma : \mathbb{G} \to \mathbb{G}$ that preserves time separation and distance between simultaneous events is necessarily a Galilean symmetry. In this one respect, the structure of $\mathbb{G}$ is harder to describe than either that of Euclidean space $\mathbb{E}$ or Minkowski space $\mathbb{M}$ (the space–time of Einstein's theory).

1.9 Historical Note

The 'chief world systems' of Galileo's title [3] were the Ptolemaic and Copernican cosmologies. The former, favoured by Aristotelian philosophers of his day, and held by the Church to be endorsed by scripture, placed the earth at rest at the centre of the universe; the latter had the sun at the centre and the earth orbiting around it. It is this book that brought Galileo to trial before the Inquisition. In the dialogue, Salviati speaks for the Copernican view of the world; the famous passage quoted in §1.2 is part of his explanation of why we are unaware of the earth's motion. Later in the book, the behaviour of the tides is held to be evidence that the earth does move—an incorrect argument, based on a false explanation of the tides. The full title was much longer, and ends '... propounding inconclusively the philosophical and physical reasons as much for one side as for the other'. Although Salviati clearly wins the argument, Galileo maintained formal neutrality; he had been led to believe that if he presented the Copernican viewpoint only as a hypothesis—a model for astronomical calculations—and did not assert the reality of the earth's motion, then he would be safe. There is an irony here since relativity theory allows no meaning to the statement 'the earth is moving in an absolute sense'.

EXERCISES

1.1. Show that the isometries of space form a group, and that the proper isometries form a subgroup.

1.2. A Galilean transformation between the coordinate systems of two inertial frames R and R' is given by

$$\begin{pmatrix} t \\ x \\ y \\ z \end{pmatrix} = \begin{pmatrix} 1 & 0 & 0 & 0 \\ v_1 & H_{11} & H_{12} & H_{13} \\ v_2 & H_{21} & H_{22} & H_{23} \\ v_3 & H_{31} & H_{32} & H_{33} \end{pmatrix} \begin{pmatrix} t' \\ x' \\ y' \\ z' \end{pmatrix} + \begin{pmatrix} c_0 \\ c_1 \\ c_2 \\ c_3 \end{pmatrix}$$

where $H \in \mathrm{SO}(3)$, the group of 3×3 orthogonal matrices with unit determinant.

(i) Show that v_1, v_2, v_3 are the components in R of the velocity of the origin of R' relative to R.

(ii) Show that the composition of two such transformations is again Galilean. How are the v entries in the composite transformation related to those in the original transformation? How does this relationship reflect the vector addition law for velocities? Find the inverse of a Galilean transformation and show that it is Galilean.

(iii) Let (t_1, x_1, y_1, z_1) and (t_2, x_2, y_2, z_2) be the coordinates relative to R of two events. Show that under Galilean transformations

(A) The time separation $t_2 - t_1$ is invariant.

(B) The spatial distance

$$\sqrt{(x_2 - x_1)^2 + (y_2 - y_1)^2 + (z_2 - z_1)^2}$$

is invariant provided that the events are simultaneous.

Explain why (B) is not true for events that are not simultaneous. Show that a transformation

$$\begin{pmatrix} t \\ x \\ y \\ z \end{pmatrix} = M \begin{pmatrix} t' \\ x' \\ y' \\ z' \end{pmatrix} + \begin{pmatrix} c_0 \\ c_1 \\ c_2 \\ c_3 \end{pmatrix}$$

with properties (A) and (B), where M is a 4×4 matrix with positive determinant, is a Galilean transformation.

1.3. The Persian and Greek armies march along a straight road at different constant speeds. They keep an eye on each other by sending back and forth scouts on foot or on horseback. The scouts travel at constant speeds, but not at the same speeds. A traveller is walking along the road between the two armies at constant speed. The Greek army sends out two scouts simultaneously, one on horseback, the other on foot. The Persian army does the same at a different time. The Greek horseman reaches the Persian army, and immediately sets out to return to the Greek army; the Persian horseman similarly reaches the Greek army and immediately sets out to return to the Persian army. The Greek foot-soldier arrives at the Persian army at the same time as the returning Persian horseman, and the Persian foot-soldier arrives at the Greek army at the same time as the returning Greek horseman. On both their outward and return journeys, the two horsemen pass each other as they pass the traveller. Show that the two foot-soldiers also pass each other as they pass the traveller.

2
Maxwell's Theory

2.1 Introduction

Newton's laws hold in all inertial frames. The formalism of classical mechanics is invariant under Galilean transformations and it is impossible to tell by observing the dynamical behaviour of particles and other bodies whether a frame of reference is at rest or in uniform motion. In the world of classical mechanics, therefore:

Principle of Relativity

There is no absolute standard of rest; only relative motion is observable.

Relativity theory takes this principle as fundamental, as a statement about the nature of space and time as much as about the properties of the Newtonian equations of motion. But if it is to be given such universal significance, then it must apply to all of physics, and not just to Newtonian dynamics. At first this seems unproblematic—it is hard to imagine that it holds at such a basic level, but not for more complex physical interactions. Nonetheless, deep problems emerge when we try to extend it to electromagnetism since Galilean invariance conflicts with Maxwell's equations.

All appears straightforward for systems involving slow-moving charges and slowly varying electric and magnetic fields. These are governed by laws that appear to be invariant under transformations between uniformly moving frames of reference. One can imagine a modern version of Galileo's ship also carrying some magnets, batteries, semiconductors, and other electrical components. Salviati's argument for relativity would seem just as compelling.

The problem arises when we include rapidly varying fields—in particular, when we consider the propagation of light. As Einstein put it, 'Maxwell's electrodynamics ..., when applied to moving bodies, leads to asymmetries which do not appear to be inherent in the phenomena' [2]. The central difficulty is that Maxwell's equations give light, along with other electromagnetic waves, a definite velocity: in empty space, it travels with the same speed in every direction, independently of the motion of the source—a fact that is incompatible with Galilean invariance. Light travelling with speed c in one frame should have speed $c + u$ in a frame moving towards the source of the light with speed u. Thus it should be possible for light to travel with any speed. Light that travels with speed c in a frame in which its source is at rest should have some other speed in a moving frame; so Galilean invariance would imply dependence of the velocity of light on the motion of the source.

In this chapter, we shall look at the basic equations of electromagnetism, Maxwell's equations and the Lorentz force law, with a view to understanding more clearly the contradiction with Galilean invariance. We shall then turn in Chapter 3 to Einstein's resolution, which is to keep the principle of relativity, but to replace the Galilean transformation by the *Lorentz transformation.*

2.2 The Unification of Electricity and Magnetism

Maxwell's crowning achievement was to write down a consistent system of partial differential equations that described the whole range of known interactions of electric and magnetic fields with moving charges. His equations unified the treatment of electricity and magnetism by revealing for the first time the full duality between the electric and magnetic fields. They have been verified over an almost unimaginable variety of physical processes, from the propagation of light over cosmological distances, through the behaviour of the magnetic fields of stars and the everyday applications in electrical engineering and laboratory experiments, down—in their quantum version—to the exchange of photons between individual electrons.

The equations remain at the centre of modern physics; but Maxwell's understanding of the nature of electric and magnetic fields, and of the role of the 'electromagnetic ether', has not survived, except in archaic and misleading terminology that still pervades the subject—'magnetic flux', 'lines of force', and so on. We shall return to this matter shortly since it is central to the apparent conflict with relativity.

2.3 Charges, Fields, and the Lorentz Force Law

The basic objects in the modern form of Maxwell's theory are

- charged particles; and
- the electric and magnetic fields $\boldsymbol{E}$ and $\boldsymbol{B}$, which are vector quantities that depend on position and time.

The charge e of a particle, which can be positive or negative, is an intrinsic quantity analogous to gravitational mass. It determines the strength of the particle's interaction with the electric and magnetic fields—as its mass determines the strength of its interaction with gravitational fields.

The interaction is in two directions. First, electric and magnetic fields exert a force on a charged particle which depends on the value of the charge, the particle's velocity, and the values of $\boldsymbol{E}$ and $\boldsymbol{B}$ at the location of the particle. The force is given by the *Lorentz force law*

$$\boldsymbol{f} = e(\boldsymbol{E} + \boldsymbol{u} \wedge \boldsymbol{B}), \tag{2.1}$$

in which e is the charge and $\boldsymbol{u}$ is the velocity. It is analogous to the gravitational force

$$\boldsymbol{f} = m\boldsymbol{g} \tag{2.2}$$

on a particle of mass m in a gravitational field $\boldsymbol{g}$. It is through the force law that an observer can, in principle, measure the electric and magnetic fields at a point, by measuring the force on a standard charge moving with known velocity.

Second, moving charges generate electric and magnetic fields. We shall not yet consider in detail the way in which they do this, beyond stating the following basic principles.

EM1

The fields depend linearly on the charges.

This means if we superimpose two distributions of charge, then the resultant $\boldsymbol{E}$ and $\boldsymbol{B}$ fields are the sums of the respective fields that the two distributions generate separately.

EM2

A stationary point charge e generates an electric field, but no magnetic field. The electric field is given by

$$\boldsymbol{E} = \frac{ke\boldsymbol{r}}{r^3} \tag{2.3}$$

where $\boldsymbol{r}$ is the position vector from the charge, $r = |\boldsymbol{r}|$, and k is a positive constant, analogous to the gravitational constant.

By combining (2.3) and (2.1) we obtain an inverse-square-law *electrostatic force*

$$\frac{kee'}{r^2} \tag{2.4}$$

between two stationary charges; unlike gravity, it is repulsive when the charges have the same sign.

EM3

A point charge moving with velocity $\boldsymbol{v}$ generates a magnetic field

$$\boldsymbol{B} = \frac{k'e\boldsymbol{v} \wedge \boldsymbol{r}}{r^3} \tag{2.5}$$

where k' is a second positive constant.

This is extrapolated from measurements of the magnetic field generated by currents flowing in electrical circuits.

The constants k and k' in EM2 and EM3 determine the strengths of electric and magnetic interactions. They are usually denoted by

$$k = \frac{1}{4\pi\epsilon_0}, \qquad k' = \frac{\mu_0}{4\pi}. \tag{2.6}$$

Charge e is measured in coulombs, $|\boldsymbol{B}|$ in teslas, and $|\boldsymbol{E}|$ in volts per metre. With other quantities in SI units,

$$\epsilon_0 = 8.9 \times 10^{-12}, \qquad \mu_0 = 1.3 \times 10^{-6}. \tag{2.7}$$

The charge of an electron is -1.6×10^{-19} coulombs; the current through an electric fire is a flow of 5–10 coulombs per second. The earth's magnetic field is about 4×10^{-5} teslas; a bar magnet's is about one tesla; there is a field of about 50 teslas on the second floor of the Clarendon Laboratory in Oxford; and the magnetic field on the surface of a neutron star is about 10^8 teslas.

Although we are more aware of gravity in everyday life, it is very much weaker than the electrostatic force—the electrostatic repulsion between two protons is a factor of 1.2×10^{36} greater than their gravitational attraction (at any separation—both forces obey the inverse square law).

Our aim is to pass from EM1–EM3 to Maxwell's equations, by replacing (2.3) and (2.5) by partial differential equations that relate the field strengths to the charge and current densities ρ and $\boldsymbol{J}$ of a continuous distribution of charge. The densities are defined as the limits

$$\rho = \lim_{V\to 0}\left(\frac{\sum e}{V}\right), \qquad \boldsymbol{J} = \lim_{V\to 0}\left(\frac{\sum e\boldsymbol{v}}{V}\right), \tag{2.8}$$

where V is a small volume containing the point, e is a charge within the volume and $\boldsymbol{v}$ is its velocity; the sums are over the charges in V and the limits are taken as the volume is shrunk (although we shall not worry too much about the precise details of the limiting process).

2.4 Stationary Distributions of Charge

We begin the task of converting the basic principles into partial differential equations by looking at the electric field of a stationary distribution of charge, where the passage to the continuous limit is made by using Gauss's theorem to restate the inverse square law.

Gauss's thereom relates the integral of the electric field over a closed surface to the total charge contained within it. For a point charge, the electric field is given by EM2:

$$\boldsymbol{E} = \frac{e\boldsymbol{r}}{4\pi\epsilon_0 r^3}\,.$$

Since $\operatorname{div}\boldsymbol{r} = 3$ and $\operatorname{grad} r = \boldsymbol{r}/r$, we have

$$\operatorname{div}(\boldsymbol{E}) = \operatorname{div}\left(\frac{e\boldsymbol{r}}{\pi\epsilon_0 r^3}\right) = \frac{e}{4\pi\epsilon_0}\left(\frac{3}{r^3} - \frac{3\boldsymbol{r}.\boldsymbol{r}}{r^5}\right) = 0$$

everywhere except at $r = 0$. Therefore, by the divergence theorem,

$$\int_{\partial V} \boldsymbol{E}.\mathrm{d}\boldsymbol{S} = 0 \tag{2.9}$$

for any closed surface ∂V bounding a volume V that does not contain the charge.

What if the volume does contain the charge? Consider the region bounded by the sphere S_R of radius R centred on the charge; S_R has outward unit normal $\boldsymbol{r}/r$. Therefore

$$\int_{S_R} \boldsymbol{E}.\mathrm{d}\boldsymbol{S} = \frac{e}{4\pi R^2 \epsilon_0}\int_{S_R} \mathrm{d}S = \frac{e}{\epsilon_0}.$$

In particular, the value of the surface integral on the left does not depend on R.

Now consider arbitrary finite volume bounded by a closed surface S. If the charge is not inside the volume, then the integral of $\boldsymbol{E}$ over S vanishes by (2.9). If it is, then we can apply (2.9) to the volume V between S and a small sphere S_R to deduce that

$$\int_S \boldsymbol{E}.\mathrm{d}\boldsymbol{S} - \int_{S_R} \boldsymbol{E}.\mathrm{d}\boldsymbol{S} = \int_{\partial V} \boldsymbol{E}.\mathrm{d}\boldsymbol{S} = 0$$

and that the integrals of $\boldsymbol{E}$ over S and S_R are the same. Therefore

$$\int_S \boldsymbol{E}.\mathrm{d}\boldsymbol{S} = \begin{cases} e/\epsilon_0 & \text{if the charge is in the volume bounded by } S \\ 0 & \text{otherwise.} \end{cases}$$

When we sum over a distribution of charges, the integral on the left picks out the total charge within S. Therefore we have Gauss's theorem.

Proposition 2.1 (Gauss)

For any closed surface ∂V bounding a volume V,

$$\int_{\partial V} \boldsymbol{E}.\mathrm{d}S = Q/\epsilon_0$$

where $\boldsymbol{E}$ is the total electric field and Q is the total charge within V.

Now we can pass to the continuous limit. Suppose that $\boldsymbol{E}$ is generated by a distribution of charges with density ρ (charge per unit volume). Then by Gauss's theorem,

$$\int_{\partial V} \boldsymbol{E}.\mathrm{d}\boldsymbol{S} = \frac{1}{\epsilon_0}\int_V \rho\,\mathrm{d}V$$

for any volume V. But then, by the divergence theorem,

$$\int_V (\operatorname{div}\boldsymbol{E} - \rho/\epsilon_0)\,\mathrm{d}V = 0\,.$$

Since this holds for any volume V, it follows that

$$\operatorname{div}\boldsymbol{E} = \rho/\epsilon_0\,. \tag{2.10}$$

By an argument in a similar spirit, we can also show that the electric field of a stationary distribution of charge is *conservative* in the sense that the total work done by the field when a charge is moved around a closed loop vanishes; that is,

$$\oint \boldsymbol{E}.\mathrm{d}\boldsymbol{s} = 0$$

for any closed path. This is equivalent to

$$\operatorname{curl} \boldsymbol{E} = 0\,, \tag{2.11}$$

since, by Stokes' theorem,

$$\oint \boldsymbol{E}.\mathrm{d}\boldsymbol{s} = \int_S \operatorname{curl} \boldsymbol{E}.\mathrm{d}\boldsymbol{S}\,,$$

where S is any surface spanning the path. This vanishes for every path and for every S if and only if (2.11) holds.

The field of a single stationary charge is conservative since

$$\boldsymbol{E} = -\operatorname{grad}\phi, \quad \text{where} \quad \phi = \frac{e}{4\pi\epsilon_0 r}\,,$$

and therefore $\operatorname{curl} \boldsymbol{E} = 0$ since the curl of a gradient vanishes identically. For a continuous distribution, $\boldsymbol{E} = -\operatorname{grad}\phi$, where

$$\phi(\boldsymbol{r}) = \frac{1}{4\pi\epsilon_0}\int_{\boldsymbol{r}'\in V} \frac{\rho(\boldsymbol{r}')}{|\boldsymbol{r}-\boldsymbol{r}'|}\,\mathrm{d}V'\,. \tag{2.12}$$

In the integral, $\boldsymbol{r}$ (the position of the point at which ϕ is evaluated) is fixed, and the integration is over the positions $\boldsymbol{r}'$ of the individual charges. In spite of the singularity at $\boldsymbol{r} = \boldsymbol{r}'$, the integral is well defined (see Exercise 3.3). So (2.11) also holds for a continuous distribution of stationary charge.

2.5 The Divergence of the Magnetic Field

We can apply the same argument that established Gauss's theorem to the magnetic field of a slow-moving charge. Here

$$\boldsymbol{B} = \frac{\mu_0 e \boldsymbol{v} \wedge \boldsymbol{r}}{4\pi r^3}\,,$$

where $\boldsymbol{r}$ is the vector from the charge to the point at which the field is measured. Since $\boldsymbol{r}/r^3 = -\operatorname{grad}(1/r)$, we have

$$\operatorname{div}\left(\boldsymbol{v} \wedge \frac{\boldsymbol{r}}{r^3}\right) = \boldsymbol{v} \wedge \operatorname{curl}\left(\operatorname{grad}\frac{1}{r}\right) = 0\,.$$

Therefore div $\boldsymbol{B} = 0$ except at $r = 0$, as in the case of the electric field. However in the magnetic case the integral of the field over a surface surrounding the charge also vanishes, since if S_R is a sphere of radius R centred on the charge, then

$$\int_{S_R} \boldsymbol{B}.\mathrm{d}\boldsymbol{S} = \frac{\mu_0 e}{4\pi} \int_{S_R} \frac{\boldsymbol{v} \wedge \boldsymbol{r}}{r^3} . \frac{\boldsymbol{r}}{r} \, \mathrm{d}S = 0 \, .$$

By the divergence theorem, the same is true for any surface surrounding the charge. We deduce that if magnetic fields are generated only by moving charges, then

$$\int_{\partial V} \boldsymbol{B}.\mathrm{d}\boldsymbol{S} = 0$$

for any volume V, and hence that

$$\operatorname{div} \boldsymbol{B} = 0 \, . \tag{2.13}$$

Of course if there were free 'magnetic poles' generating magnetic fields in the same way that charges generate electric fields, then this would not hold; there would be a 'magnetic pole density' on the right-hand side, by analogy with the charge density in (2.10).

2.6 Inconsistency with Galilean Relativity

Our central concern is the compatibility of the laws of electromagnetism with the principle of relativity. As Einstein observed, simple electromagnetic interactions do indeed depend only on relative motion; the current induced in a conductor moving through the field of magnet is the same as that generated in a stationary conductor when a magnet is moved past it with the same relative velocity [2]. Unfortunately this symmetry is not reflected in our basic principles. We very quickly come up against contradictions if we assume that they hold in every inertial frame of reference.

One emerges as follows. An observer O can measure the values of $\boldsymbol{B}$ and $\boldsymbol{E}$ at a point by measuring the force on a particle of standard charge, which is related to the velocity $\boldsymbol{v}$ of the charge by the Lorentz force law,

$$\boldsymbol{f} = e(\boldsymbol{E} + \boldsymbol{v} \wedge \boldsymbol{B}).$$

A second observer O' moving relative to the first with velocity $\boldsymbol{v}$ will see the same force, but now acting on a particle at rest. He will therefore measure the electric field to be $\boldsymbol{E}' = \boldsymbol{f}/e$. We conclude that an observer moving with velocity $\boldsymbol{v}$ through a magnetic field $\boldsymbol{B}$ and an electric field $\boldsymbol{E}$ should see an electric field

$$\boldsymbol{E}' = \boldsymbol{E} + \boldsymbol{v} \wedge \boldsymbol{B}. \tag{2.14}$$

By interchanging the roles of the two observers, we should also have

$$\boldsymbol{E} = \boldsymbol{E}' - \boldsymbol{v} \wedge \boldsymbol{B}' \tag{2.15}$$

where $\boldsymbol{B}'$ is the magnetic field measured by the second observer. If both are to hold, then $\boldsymbol{B} - \boldsymbol{B}'$ must be a scalar multiple of $\boldsymbol{v}$.

But this is incompatible with EM3; if the fields are those of a point charge at rest relative to the first observer, then $\boldsymbol{E}$ is given by (2.3), and

$$\boldsymbol{B} = 0\,.$$

On the other hand, the second observer sees the field of a point charge moving with velocity $-\boldsymbol{v}$. Therefore

$$\boldsymbol{B}' = -\frac{\mu_0 e \boldsymbol{v} \wedge \boldsymbol{r}}{4\pi r^3}\,.$$

So $\boldsymbol{B} - \boldsymbol{B}'$ is orthogonal to $\boldsymbol{v}$, not parallel to it.

This conspicuous paradox is resolved, in part, by the realization that EM3 is not exact; it holds only when the velocities are small enough for the magnetic force between two particles to be negligible in comparison with the electrostatic force. If v is a typical velocity, then the condition is that $v^2\mu_0$ should be much less than $1/\epsilon_0$. That is, the velocities involved should be much less than

$$c = \frac{1}{\sqrt{\epsilon_0 \mu_0}} = 3 \times 10^8\,\mathrm{m/s}\,.$$

We shall see later that c is the velocity of light.

2.7 The Limits of Galilean Invariance

Our basic principles EM1–EM3 must now be seen to be approximations—they describe the interactions of particles and fields when the particles are moving relative to each other at speeds much less than that of light. To emphasize that we cannot expect, in particular, EM3 to hold for particles moving at speeds comparable with c, we must replace it by

EM3′

A charge moving with velocity $\boldsymbol{v}$, where $v \ll c$, generates a magnetic field

$$\boldsymbol{B} = \frac{\mu_0 e \boldsymbol{v} \wedge \boldsymbol{r}}{4\pi r^3} + O(v^2/c^2)\,. \tag{2.16}$$

The magnetic field of a system of charges in general motion satisfies

$$\operatorname{div} \boldsymbol{B} = 0\,. \tag{2.17}$$

In the second part, we have retained (2.17) as a differential form of the statement that there are no free magnetic poles; the magnetic field is generated only by the motion of the charges. With this change, the theory is consistent with the principle of relativity, provided that we ignore terms of order v^2/c^2. The substitution of EM3′ for EM3 resolves the conspicuous paradox; the symmetry noted by Einstein between the current generated by the motion of the conductor in a magnetic field and by the motion of a magnet past a conductor is explained, provided that the velocities are much less than that of light.

The central problem remains however; the equations of electromagnetism are not invariant under a Galilean transformation with velocity comparable to c. The paradox is still there, but it is more subtle than it appeared to be at first. There are three possible ways out: either the noninvariance is real and has observable effects (necessarily of order v^2/c^2 or smaller); or Maxwell's theory is wrong; or the Galilean transformation is wrong. Disconcertingly, it is the last path that we shall take. But that is to jump ahead in the story. We must first complete the derivation of Maxwell's equations.

2.8 Faraday's Law of Induction

The magnetic field of a slow-moving charge will always be small in relation to its electric field (even when we replace $\boldsymbol{B}$ by $c\boldsymbol{B}$ to put it into the same units as $\boldsymbol{E}$). The magnetic fields generated by currents in electrical circuits are not, however, dominated by large electric fields. This is because the currents are created by the flow, at slow velocity, of electrons, while overall the matter in the wire is roughly electrically neutral, with the electric fields of the positively charged nuclei and negatively charged electrons cancelling.

This is the physical context to keep in mind in the following deduction of Faraday's law of induction from Galilean invariance for velocities much less than c. The law relates the electromotive force or 'voltage' around an electrical circuit to the rate of change of the magnetic field $\boldsymbol{B}$ over a surface spanning the circuit. In its differential form, the law becomes one of Maxwell's equations.

Suppose first that the fields are generated by charges all moving relative to a given inertial frame of reference R with the same velocity $\boldsymbol{v}$. Then in a second frame R' moving relative to R with velocity $\boldsymbol{v}$, there is a stationary distribution of charge. If the velocity is much less than that of light, then the electric field $\boldsymbol{E}'$ measured in R' is related to the electric and magnetic $\boldsymbol{E}$ and $\boldsymbol{B}$ measured in R by

$$\boldsymbol{E}' = \boldsymbol{E} + \boldsymbol{v} \wedge \boldsymbol{B} .$$

Since the field measured in R' is that of a stationary distribution of charge, we have

$$\operatorname{curl} \boldsymbol{E}' = 0\,.$$

In R, the charges are all moving with velocity $\boldsymbol{v}$, so their configuration looks exactly the same from the point $\boldsymbol{r}$ at time t as it does from the point $\boldsymbol{r} + \boldsymbol{v}\tau$ at time $t + \tau$. Therefore

$$\boldsymbol{B}(\boldsymbol{r} + \boldsymbol{v}\tau, t + \tau) = \boldsymbol{B}(\boldsymbol{r}, t), \qquad \boldsymbol{E}(\boldsymbol{r} + \boldsymbol{v}\tau, t + \tau) = \boldsymbol{E}(\boldsymbol{r}, t)\,,$$

and hence by taking derivatives with respect to τ at $\tau = 0$,

$$\boldsymbol{v}\,.\operatorname{grad} \boldsymbol{B} + \frac{\partial \boldsymbol{B}}{\partial t} = 0, \qquad \boldsymbol{v}\,.\operatorname{grad} \boldsymbol{E} + \frac{\partial \boldsymbol{E}}{\partial t} = 0\,. \tag{2.18}$$

So we must have

$$\begin{aligned} 0 &= \operatorname{curl} \boldsymbol{E}' \\ &= \operatorname{curl} \boldsymbol{E} + \operatorname{curl}(\boldsymbol{v} \wedge \boldsymbol{B}) \\ &= \operatorname{curl} \boldsymbol{E} + \boldsymbol{v} \operatorname{div} \boldsymbol{B} - \boldsymbol{v}\,.\operatorname{grad} \boldsymbol{B} \\ &= \operatorname{curl} \boldsymbol{E} + \frac{\partial \boldsymbol{B}}{\partial t} \end{aligned} \tag{2.19}$$

since $\operatorname{div} \boldsymbol{B} = 0$. It follows that

$$\operatorname{curl} \boldsymbol{E} + \frac{\partial \boldsymbol{B}}{\partial t} = 0\,. \tag{2.20}$$

Equation (2.20) is linear in $\boldsymbol{B}$ and $\boldsymbol{E}$; so by adding the magnetic and electric fields fields of different streams of charges moving relative to R with different velocities, we deduce that it holds generally for the electric and magnetic fields generated by moving charges.

Equation (2.20) encodes Faraday's law of *electromagnetic induction*, which describes how changing magnetic fields can generate currents. In the static case

$$\frac{\partial \boldsymbol{B}}{\partial t} = 0$$

and the equation reduces to $\operatorname{curl} \boldsymbol{E} = 0$—the condition that the electrostatic field should be conservative; that is, it should do no net work when a charge is moved around a closed loop.

More generally, consider a wire loop in the shape of a closed curve γ. Let S be a fixed surface spanning γ. Then we can deduce from Equation (2.20) that

$$\begin{aligned} \oint_\gamma \boldsymbol{E}\,.\,\mathrm{d}\boldsymbol{s} &= \int_S \operatorname{curl} \boldsymbol{E}\,.\,\mathrm{d}\boldsymbol{S} \\ &= -\int_S \frac{\partial \boldsymbol{B}}{\partial t}\,.\,\mathrm{d}\boldsymbol{S} \\ &= -\frac{\mathrm{d}}{\mathrm{d}t}\int_S (\boldsymbol{B}\,.\,\mathrm{d}\boldsymbol{S})\,. \end{aligned} \tag{2.21}$$

If the magnetic field is varying, so that the integral of $\boldsymbol{B}$ over S is not constant, then the integral of $\boldsymbol{E}$ around the loop will not be zero. There will be a nonzero electric field along the wire, which will exert a force on the electrons in the wire and cause a current to flow.

The quantity

$$\oint \boldsymbol{E} \, . \, \mathrm{d}\boldsymbol{s} \, ,$$

which is measured in volts, is the work done by the electric field when a unit charge makes one circuit of the wire. It is called the *electromotive force* around the circuit. The integral is the *magnetic flux* linking the circuit. The relationship (2.21) between electromotive force and rate of change of magnetic flux is *Faraday's law.*

2.9 The Field of Charges in Uniform Motion

We can extract another of Maxwell's equations from this argument. By EM3′, a single charge e with velocity $\boldsymbol{v}$ generates an electric field $\boldsymbol{E}$ and a magnetic field

$$\boldsymbol{B} = \frac{\mu_0 e \boldsymbol{v} \wedge \boldsymbol{r}}{4\pi r^3} + O(v^2/c^2) \, ,$$

where $\boldsymbol{r}$ is the vector from the charge to the point at which the field is measured. In the frame of reference R' in which the charge is at rest, its electric field is

$$\boldsymbol{E}' = \frac{e\boldsymbol{r}}{4\pi\epsilon_0 r^3} \, .$$

In the frame in which it is moving with velocity $\boldsymbol{v}$, $\boldsymbol{E} = \boldsymbol{E}' + O(v/c)$. Therefore

$$c\boldsymbol{B} = \frac{\boldsymbol{v} \wedge \boldsymbol{E}'}{c} = \frac{\boldsymbol{v} \wedge \boldsymbol{E}}{c} + O\left(\frac{v^2}{c^2}\right) .$$

By taking the curl of both sides, and dropping terms of order v^2/c^2,

$$\begin{aligned} \operatorname{curl}(c\boldsymbol{B}) &= \operatorname{curl}\left(\frac{\boldsymbol{v} \wedge \boldsymbol{E}}{c}\right) \\ &= \frac{1}{c}\Big(\boldsymbol{v} \operatorname{div} \boldsymbol{E} - \boldsymbol{v} \, . \operatorname{grad} \boldsymbol{E}\Big) . \end{aligned}$$

But

$$\operatorname{div} \boldsymbol{E} = \rho/\epsilon_0 \qquad \text{and} \qquad \boldsymbol{v} \, . \operatorname{grad} \boldsymbol{E} = -\frac{\partial \boldsymbol{E}}{\partial t} \, ,$$

by (2.18). Therefore

$$\operatorname{curl}(c\boldsymbol{B}) - \frac{1}{c}\frac{\partial \boldsymbol{E}}{\partial t} = \frac{1}{c\epsilon_0}\boldsymbol{J} = c\mu_0 \boldsymbol{J}$$

where $\boldsymbol{J} = \rho \boldsymbol{v}$. By summing over the separate particle velocities, we conclude that

$$\operatorname{curl} \boldsymbol{B} - \frac{1}{c^2}\frac{\partial \boldsymbol{E}}{\partial t} = \mu_0 \boldsymbol{J}$$

holds for an arbitrary distribution of charges, provided that their velocities are much less than that of light.

2.10 Maxwell's Equations

The basic principles, together with the assumption of Galilean invariance for velocities much less than that of light, have allowed us to deduce that the electric and magnetic fields generated by a continuous distribution of moving charges in otherwise empty space satisfy

$$\operatorname{div} \boldsymbol{E} = \frac{\rho}{\epsilon_0} \tag{2.22}$$

$$\operatorname{div} \boldsymbol{B} = 0 \tag{2.23}$$

$$\operatorname{curl} \boldsymbol{B} - \frac{1}{c^2}\frac{\partial \boldsymbol{E}}{\partial t} = \mu_0 \boldsymbol{J} \tag{2.24}$$

$$\operatorname{curl} \boldsymbol{E} + \frac{\partial \boldsymbol{B}}{\partial t} = 0, \tag{2.25}$$

where ρ is the charge density, $\boldsymbol{J}$ is the current density, and $c^2 = 1/\epsilon_0\mu_0$. These are *Maxwell's equations*, the basis of modern electrodynamics. Together with the Lorentz force law, they describe the dynamics of charges and electromagnetic fields.

We have arrived at them by considering how basic electromagnetic processes appear in moving frames of reference—an unsatisfactory route because we have seen on the way that the principles on which we based the derivation are incompatible with Galilean invariance for velocities comparable with that of light. Maxwell derived them by analysing an elaborate mechanical model of electric and magnetic fields—as displacements in the *luminiferous ether*. That is also unsatisfactory because the model has long been abandoned. The reason that they are accepted today as the basis of theoretical and practical applications of electromagnetism has little to do with either argument. It is first that they are self-consistent, and second that they describe the behaviour of real fields with unreasonable accuracy.

2.11 The Continuity Equation

It is not immediately obvious that the equations *are* self-consistent. Given ρ and $\boldsymbol{J}$ as functions of the coordinates and time, Maxwell's equations are two scalar and two vector equations in the unknown components of $\boldsymbol{E}$ and $\boldsymbol{B}$. That is, a total of eight equations for six unknowns—more equations than unknowns. Therefore it is possible that they are in fact inconsistent.

If we take the divergence of the fourth equation (2.25), then we obtain

$$\frac{\partial}{\partial t}\Big(\operatorname{div}\boldsymbol{B}\Big) = 0\,,$$

which is consistent with the second; so no problem arises here. However, by taking the divergence of the third equation and substituting from the first, we get

$$\begin{aligned} 0 &= \operatorname{div}\operatorname{curl}\boldsymbol{B} \\ &= \frac{1}{c^2}\frac{\partial}{\partial t}\Big(\operatorname{div}\boldsymbol{E}\Big) + \mu_0 \operatorname{div}\boldsymbol{J} \\ &= \mu_0\left(\frac{\partial\rho}{\partial t} + \operatorname{div}\boldsymbol{J}\right). \end{aligned}$$

This gives a contradiction unless

$$\frac{\partial\rho}{\partial t} + \operatorname{div}\boldsymbol{J} = 0\,. \tag{2.26}$$

So the choice of ρ and $\boldsymbol{J}$ is not unconstrained; they must be related by the *continuity equation* (2.26). This holds for physically reasonable distributions of charge; it is a differential form of the statement that charges are neither created nor destroyed.

2.12 Conservation of Charge

To see the connection between the continuity equation and charge conservation, let us look at the total charge within a fixed V bounded by a surface S. If charge is conserved, then any increase or decrease in a short period of time must be exactly balanced by an inflow or outflow of charge across S.

Consider a small element $\mathrm{d}\boldsymbol{S}$ of S with outward unit normal and consider all the particles that have a particular charge e and a particular velocity $\boldsymbol{v}$ at time t. Suppose that there are σ of these per unit volume (σ is a function of position). Those that cross the surface element between t and $t + \delta t$ are those that at time t lie in the region of volume

$$|\boldsymbol{v} \,.\, \boldsymbol{n} \, \mathrm{d}S \, \delta t|$$

shown in Figure 2.1. They contribute $e\sigma\boldsymbol{v} \,.\, \mathrm{d}\boldsymbol{S} \, \delta t$ to the outflow of charge

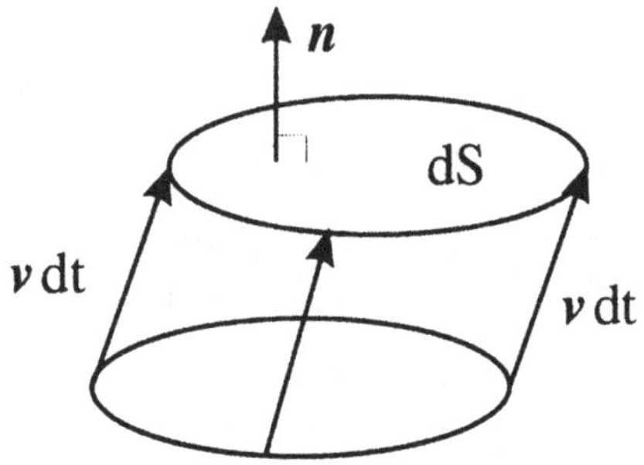

Figure 2.1 The outflow through a surface element

through the surface element. But the value of $\boldsymbol{J}$ at the surface element is the sum of $e\sigma\boldsymbol{v}$ over all possible values of $\boldsymbol{v}$ and e. By summing over $\boldsymbol{v}$, e, and the elements of the surface, therefore, and by passing to the limit of a continuous distribution, the total rate of outflow is

$$\int_S \boldsymbol{J} \,.\, \mathrm{d}\boldsymbol{S} \,.$$

Charge conservation implies that the rate of outflow should be equal to the rate of decrease in the total charge within V. That is,

$$\frac{\mathrm{d}}{\mathrm{d}t} \int_V \rho \, \mathrm{d}V + \int_S \boldsymbol{J} \,.\, \mathrm{d}\boldsymbol{S} = 0 \,. \tag{2.27}$$

By differentiating the first term under the integral sign and by applying the divergence theorem to the second integral,

$$\int_V \left(\frac{\partial \rho}{\partial t} + \operatorname{div} \boldsymbol{J} \right) \mathrm{d}V = 0 \,. \tag{2.28}$$

If this is to hold for any choice of V, then ρ and $\boldsymbol{J}$ must satisfy the continuity equation. Conversely, the continuity equation implies charge conservation.

2.13 Historical Note

At the end of the eighteenth century, four types of electromagnetic phenomena were known, but not the connections between them.

- *Magnetism*; the word derives from the Greek for 'stone from Magnesia'.
- *Static electricity*, produced by rubbing amber with fur; the word 'electricity' derives from the Greek for 'amber'.
- *Light*.
- *Galvanism* or 'animal electricity'—the electricity produced by batteries, discovered by Luigi Galvani.

The construction of a unified theory was a slow and painful business. It was hindered by attempts, which seem bizarre in retrospect, to understand electromagnetism in terms of underlying mechanical models involving such inventions as 'electric fluids' and 'magnetic vortices'. We can see the legacy of this period, which ended with Einstein's work in 1905, in the misleading and archaic terms that still survive in modern terminology: 'magnetic flux', 'lines of force', 'electric displacement', and so on.

Maxwell's contribution was decisive, although much of what is here called 'Maxwell's theory' is due to his successors (Lorentz, Hertz, Einstein, ...); and, as we shall see, a key element in Maxwell's own description of electromagnetism—the 'electromagnetic ether', an all-pervasive medium which was supposed to transmit electromagnetic waves—was thrown out by Einstein.

A rough chronology is as follows.

1800 Volta demonstrated the connection between Galvanism and static electricity.

1820 Oersted showed that the current from a battery generates a force on a magnet.

1822 Ampère suggested that light was a wave motion in a 'luminiferous ether' made up of two types of electric fluid. In the same year, Galileo's *Dialogue concerning the two chief world systems* [3] was removed from the index of prohibited books.

1831 Faraday showed that moving magnets can induce currents.

1846 Faraday suggested that light is a vibration in magnetic lines of force.

1863 Maxwell published the equations that describe the dynamics of electric and magnetic fields.

1905 Einstein's paper *On the electrodynamics of moving bodies* [2].

EXERCISES

2.1. Show that if $\boldsymbol{E}$ and $\boldsymbol{B}$ satisfy Maxwell's equations with $\rho = 0 = \boldsymbol{J}$, then so do

$$\boldsymbol{E}' = \cos\alpha\,\boldsymbol{E} - \sin\alpha\, c\boldsymbol{B}, \qquad \boldsymbol{B}' = c^{-1}\sin\alpha\,\boldsymbol{E} + \cos\alpha\,\boldsymbol{B},$$

for any constant α (this transformation is called a *duality rotation*).

2.2. Let

$$\boldsymbol{B} = \begin{cases} \dfrac{I\mu_0(-y\boldsymbol{i} + x\boldsymbol{j})}{2\pi a^2} & \text{if } x^2 + y^2 < a^2 \\[2ex] \dfrac{I\mu_0(-y\boldsymbol{i} + x\boldsymbol{j})}{2\pi(x^2 + y^2)} & \text{if } x^2 + y^2 \geq a^2, \end{cases}$$

where I is constant. Show that

$$\operatorname{curl}\boldsymbol{B} = \begin{cases} \dfrac{\mu_0 I\boldsymbol{k}}{\pi a^2} & \text{if } x^2 + y^2 < a^2 \\ 0 & \text{if } x^2 + y^2 \geq a^2. \end{cases}$$

[This gives the magnetic field generated by a current I in a long straight wire.]

3
The Propagation of Light

3.1 The Displacement Current

The electric and magnetic fields of a distribution of charges with density ρ and current density $\boldsymbol{J}$ in otherwise empty space are governed by Maxwell's equations

$$\begin{aligned}
\operatorname{div} \boldsymbol{E} &= \frac{\rho}{\epsilon_0} \\
\operatorname{div} \boldsymbol{B} &= 0 \\
\operatorname{curl} \boldsymbol{B} - \frac{1}{c^2}\frac{\partial \boldsymbol{E}}{\partial t} &= \mu_0 \boldsymbol{J} \\
\operatorname{curl} \boldsymbol{E} + \frac{\partial \boldsymbol{B}}{\partial t} &= 0 .
\end{aligned}$$

These are consistent with each other provided that the continuity equation

$$\frac{\partial \rho}{\partial t} + \operatorname{div} \boldsymbol{J} = 0$$

holds. We think of ρ and $\boldsymbol{J}$ as *generating* the fields $\boldsymbol{E}$ and $\boldsymbol{B}$, and refer to them as the *sources.*

We derived the equations from basic assumptions that are consistent with Galileo's relativity principle only if we ignore second- and higher-order terms in v/c, where v is the velocity of a typical charge. So it might seem reasonable to assume that they give only an approximate description of the behaviour of real fields and that if we want to understand the behaviour of very rapidly changing fields, then we should first replace them by some other as yet unknown system of equations, which would be compatible with Galilean invariance. The remarkable fact, however, is that Maxwell's equations do indeed accurately reflect the real behaviour of rapidly varying fields and fast-moving charges, even when we drop the restriction on v/c—indeed they reflect it with almost unreasonable effectiveness. In particular, they correctly determine the propagation of electromagnetic waves—light, radio waves, X-rays, and so forth.

Maxwell arrived at his equations through a different line of reasoning. It is not altogether straightforward to reconstruct his argument since his point of view shifted as he worked out his ideas, first in the context of an elaborate model involving elastic properties of the ether, and then in their presentation in his *Dynamical theory of the electromagnetic field* [11], where the equations appear within a free-standing mathematical framework. Although he took the ether to be real, he was clear that the more intricate elements of his earlier model were to be understood in terms of *analogies* rather than as descriptions of what really happens when fields change and charges interact.

His key step can be explained by writing the third equation in the form

$$\operatorname{curl} \boldsymbol{B} = \mu_0 \left(\boldsymbol{J} + \epsilon_0 \frac{\partial \boldsymbol{E}}{\partial t} \right) , \tag{3.1}$$

so that it can be read as an equation for an unknown magnetic field $\boldsymbol{B}$ in terms of a known current distribution $\boldsymbol{J}$ and electric field $\boldsymbol{E}$. When $\boldsymbol{E}$ and $\boldsymbol{J}$ are independent of t, it reduces to

$$\operatorname{curl} \boldsymbol{B} = \mu_0 \boldsymbol{J} ,$$

which determines the magnetic field of a steady current, in a way that was already familiar to Maxwell's contemporaries. But his second term on the right-hand side of (3.1) was new; it adds to $\boldsymbol{J}$ the so-called vacuum *displacement current*

$$\epsilon_0 \frac{\partial \boldsymbol{E}}{\partial t} .$$

The name comes from an analogy with the behaviour of charges in an insulating material. Here no steady current can flow, but the distribution of charges within the material is distorted by an external electric field. When the field changes, the distortion also changes, and the result appears as a current—the displacement current—which flows during the period of change. Maxwell's central insight was that the same term should be present even in empty space. The consequence was profound; it allowed him to explain the propagation of light as an electromagnetic phenomenon.

Our route to Maxwell's equations was not the historical one, and Maxwell's analogies are certainly not very helpful to us in thinking about the relativistic aspects of electromagnetism. But if we are to understand how the equations explain the behaviour of electromagnetic waves, then we must first take a step as bold as Maxwell's and explore the consequences of accepting his equations without restriction on v/c. We shall put aside for the moment the conflict with relativity.

3.2 The Source-free Equations

In a region of empty space, away from the charges generating the electric and magnetic fields, we have $\rho = 0 = \boldsymbol{J}$, and Maxwell's equations reduce to

$$\operatorname{div} \boldsymbol{E} = 0 \tag{3.2}$$

$$\operatorname{div} \boldsymbol{B} = 0 \tag{3.3}$$

$$\operatorname{curl} \boldsymbol{B} - \frac{1}{c^2}\frac{\partial \boldsymbol{E}}{\partial t} = 0 \tag{3.4}$$

$$\operatorname{curl} \boldsymbol{E} + \frac{\partial \boldsymbol{B}}{\partial t} = 0\,, \tag{3.5}$$

where $c = 1/\sqrt{\epsilon_0 \mu_0}$. By taking the curl of Equation (3.4) and by substituting from Equations (3.3) and (3.5), we obtain, with the help of Equation (B.7),

$$\begin{aligned} 0 &= \operatorname{grad}(\operatorname{div} \boldsymbol{B}) - \nabla^2 \boldsymbol{B} - \frac{1}{c^2} \operatorname{curl}\left(\frac{\partial \boldsymbol{E}}{\partial t}\right) \\ &= -\nabla^2 \boldsymbol{B} - \frac{1}{c^2}\frac{\partial}{\partial t}(\operatorname{curl} \boldsymbol{E}) \\ &= -\nabla^2 \boldsymbol{B} + \frac{1}{c^2}\frac{\partial^2 \boldsymbol{B}}{\partial t^2}\,. \end{aligned} \tag{3.6}$$

Therefore the three components of $\boldsymbol{B}$ in empty space satisfy the (scalar) *wave equation*

$$\Box u = 0\,.$$

Here $\Box$ is the *d'Alembertian* operator, defined by

$$\Box = \frac{1}{c^2}\frac{\partial^2}{\partial t^2} - \nabla^2 = \frac{1}{c^2}\frac{\partial^2}{\partial t^2} - \frac{\partial^2}{\partial x^2} - \frac{\partial^2}{\partial y^2} - \frac{\partial^2}{\partial z^2}\,.$$

By taking the curl of Equation (3.5), we also obtain $\Box \boldsymbol{E} = 0$.

3.3 The Wave Equation

The scalar wave equation

$$\frac{1}{c^2}\frac{\partial^2 u}{\partial t^2} - \frac{\partial^2 u}{\partial x^2} - \frac{\partial^2 u}{\partial y^2} - \frac{\partial^2 u}{\partial z^2} = 0 \tag{3.7}$$

governs the propagation of waves in three-dimensional space. It generalizes the one-dimensional equation

$$\frac{1}{c^2}\frac{\partial^2 u}{\partial t^2} - \frac{\partial^2 u}{\partial x^2} = 0 \tag{3.8}$$

(in both cases, c is the velocity of the waves).

In one dimension, the behaviour of u is simple to understand. For any function f of a single variable,

$$u(x,t) = f(x - ct) \qquad \text{and} \qquad u(x,t) = f(x + ct) \tag{3.9}$$

are solutions (we assume that f is sufficiently smooth to justify the calculations that follow). The first represents a wave profile given by the graph of f, travelling with velocity c in the direction of increasing x; and the second represents the same wave profile travelling with the same speed, but in the opposite direction. The general solution is a superposition of two such travelling waves:

$$u = f_1(x - ct) + f_2(x + ct)\,.$$

In three dimensions, travelling waves can move in any direction. For any constant unit vector $\boldsymbol{e}$ and any suitable function f of a single variable,

$$u = f(\boldsymbol{e}\,.\,\boldsymbol{r} - ct)$$

is a solution of (3.7). One can see this directly by using

$$\nabla^2 u = \boldsymbol{e}\,.\,\boldsymbol{e}\, f'' = f'' \qquad \text{and} \qquad u_{tt} = c^2 f''\,.$$

Such a solution is called a *plane-fronted wave.* At a particular time, u is constant on the planes orthogonal to $\boldsymbol{e}$ given by

$$\boldsymbol{e}.\boldsymbol{r} = \text{constant}\,.$$

As t increases, these plane *wave fronts* progress in the direction of $\boldsymbol{e}$ with speed c.

To build up the general solution, we have to integrate over all directions—we cannot write the solution in the same simple way as in one dimension as a superposition of two waves moving in opposite directions.

The one-dimensional equation (3.8) has special travelling wave solutions called *harmonic waves*, of the form

$$u = A\cos\left[\frac{\omega}{c}(ct - x) + \epsilon\right] \tag{3.10}$$

where the *amplitude* A, the *(angular) frequency* $\omega > 0$, and the *phase* ϵ are constants. The graph of u at fixed t is a cosine curve, which moves to the right with constant velocity c as t increases. It follows from the theory of the Fourier transform that every solution of (3.9) can be written as a 'sum'—in fact, an integral—of harmonic waves.

This result does generalize to three dimensions. A *real harmonic wave* is a solution of Equation (3.7) of the form

$$u = \alpha\cos\Omega + \beta\sin\Omega\,, \tag{3.11}$$

where

$$\Omega = \frac{\omega}{c}(ct - \boldsymbol{r}\,.\,\boldsymbol{e}), \qquad \boldsymbol{e}\,.\,\boldsymbol{e} = 1\,, \tag{3.12}$$

with $\omega > 0$, α, β and $\boldsymbol{e}$ constant; ω is the frequency and $\boldsymbol{e}$ is a unit vector that gives the direction of propagation (adding τ to t and $c\tau\boldsymbol{e}$ to $\boldsymbol{r}$ leaves u unchanged). Again it follows by Fourier analysis that every solution of Equation (3.7) is a combination of harmonic waves.

3.4 Monochromatic Plane Waves

The fact that $\boldsymbol{E}$ and $\boldsymbol{B}$ are vector-valued solutions of the wave equation in empty space suggests that we look for solutions of Maxwell's equations in which

$$\boldsymbol{E} = \boldsymbol{\alpha}\cos\Omega + \boldsymbol{\beta}\sin\Omega\,, \tag{3.13}$$

where $\boldsymbol{\alpha}$, $\boldsymbol{\beta}$, are constant vectors and Ω is again given by (3.12). This satisfies the wave equation, but for a general choice of the constants, it will not be possible to find $\boldsymbol{B}$ such that Equations (3.2)–(3.5) also hold.

By taking the divergence of Equation (3.13), we obtain

$$\operatorname{div}\boldsymbol{E} = \frac{\omega}{c}\Big(\boldsymbol{e}\,.\,\boldsymbol{\alpha}\sin\Omega - \boldsymbol{e}\,.\,\boldsymbol{\beta}\cos\Omega\Big)\,, \tag{3.14}$$

For Equation (3.2) to hold, therefore, we must choose $\boldsymbol{\alpha}$ and $\boldsymbol{\beta}$ orthogonal to $\boldsymbol{e}$. For Equation (3.5) to hold, we must find $\boldsymbol{B}$ such that

$$\operatorname{curl}\boldsymbol{E} = \frac{\omega}{c}\Big(\boldsymbol{e}\wedge\boldsymbol{\alpha}\sin\Omega - \boldsymbol{e}\wedge\boldsymbol{\beta}\cos\Omega\Big) = -\frac{\partial\boldsymbol{B}}{\partial t}\,. \tag{3.15}$$

A possible choice is

$$B = \frac{e \wedge E}{c} = \frac{1}{c}\Big(e \wedge \alpha \cos \Omega + e \wedge \beta \sin \Omega\Big) \tag{3.16}$$

and it is not hard to see that E and B then satisfy (3.3) and (3.4) as well.

Exercise 3.1

Show that with these choices, E and B satisfy Maxwell's equations, without sources.

Definition 3.1

A *monochromatic electromagnetic plane wave* in vacuo of angular frequency ω is a solution of Maxwell's equations of the form

$$E = \alpha \cos \Omega + \beta \sin \Omega, \qquad B = \frac{e \wedge E}{c} \tag{3.17}$$

where

$$\Omega = \omega\left(t - \frac{r \,.\, e}{c}\right),$$

and α, β, e, $\omega > 0$ are constant, with $e \,.\, e = 1$ and $e \,.\, \alpha = 0 = e \,.\, \beta$.

Note that such waves are *transverse* in the sense that E and B are orthogonal to the direction of propagation. The definition E can be written more concisely in the form

$$E = \mathrm{Re}\left[(\alpha + \mathrm{i}\beta)\mathrm{e}^{-\mathrm{i}\Omega}\right]. \tag{3.18}$$

It is again an exercise in Fourier analysis to show every solution in empty space is a combination of monochromatic plane waves.

At the heart of Maxwell's theory was the idea that a light wave with definite frequency or colour is represented by a monochromatic plane solution of his equations.

3.5 Polarization

Consider a general monochromatic plane wave. At a fixed point in space, the electric field sweeps out an ellipse in the plane spanned by α and β perpendicular to e. At the origin, for example,

$$E = \alpha \cos(\omega t) + \beta \sin(\omega t)\,. \tag{3.19}$$

The vector cB sweeps out the same ellipse, but rotated through $\pi/2$ about e.

Exercise 3.2

Show that as t varies in (3.19), $\boldsymbol{E}$ sweeps out an ellipse.

Two special cases arise when $\boldsymbol{\alpha}$ and $\boldsymbol{\beta}$ are proportional, so that the ellipse degenerates to a line, and when $\boldsymbol{\alpha}\,.\,\boldsymbol{\alpha} = \boldsymbol{\beta}\,.\,\boldsymbol{\beta}$, $\boldsymbol{\alpha}\,.\,\boldsymbol{\beta} = 0$, so that the ellipse becomes a circle.

Definition 3.2

A plane wave has *plane* or *linear* polarization if $\boldsymbol{\alpha}$ and $\boldsymbol{\beta}$ are proportional. It has *circular* polarization if $\boldsymbol{\alpha}\,.\,\boldsymbol{\alpha} = \boldsymbol{\beta}\,.\,\boldsymbol{\beta}$, $\boldsymbol{\alpha}\,.\,\boldsymbol{\beta} = 0$.

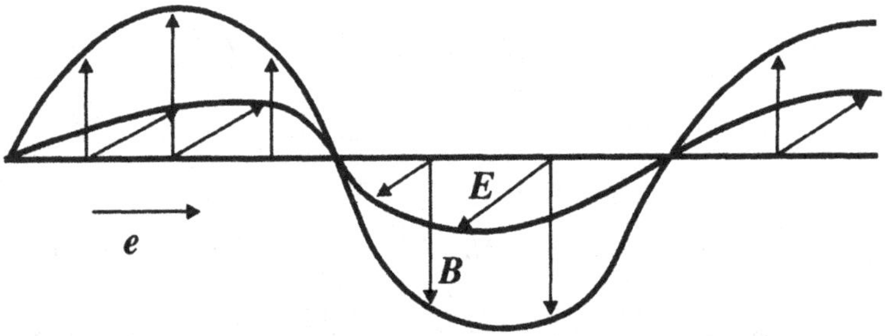

Figure 3.1 Linear polarization

Figure 3.1 shows a monochromatic linearly polarized wave. Polarizing sunglasses are more transparent to linearly polarized waves with a particular orientation of $\boldsymbol{\alpha}$ than to those with the orthogonal orientation of $\boldsymbol{\alpha}$.

A circularly polarized wave is right- or left-handed as

$$\boldsymbol{\beta} = \boldsymbol{e} \wedge \boldsymbol{\alpha} \qquad \text{or} \qquad \boldsymbol{\beta} = -\boldsymbol{e} \wedge \boldsymbol{\alpha}\,.$$

Note that a circularly polarized wave is a superposition of two linearly polarized waves which have orthogonal polarizations (directions of $\boldsymbol{\alpha}$) and are exactly out of phase.

3.6 Potentials

For every solution of Maxwell's equations *in vacuo*, the components of $\boldsymbol{E}$ and $\boldsymbol{B}$ satisfy the three-dimensional wave equation; but the converse is not true.

That is, it is not true in general that if

$$\Box \boldsymbol{B} = 0, \qquad \Box \boldsymbol{E} = 0,$$

then $\boldsymbol{E}$ and $\boldsymbol{B}$ satisfy Maxwell's equations. For this to happen, the divergence of both must vanish; and they must be related by (3.4) and (3.5). These additional constraints are somewhat simpler to handle if we work not with the fields themselves, but with auxiliary quantities called *potentials.*

The definition of the potentials depends on standard integrability conditions from vector calculus, which are proved in Appendix B. Suppose that $\boldsymbol{v}$ is a vector field, which may depend on time. It is shown in the appendix that if $\operatorname{curl} \boldsymbol{v} = 0$, then there exists a function ϕ such that

$$\boldsymbol{v} = \operatorname{grad} \phi; \tag{3.20}$$

or that if $\operatorname{div} \boldsymbol{v} = 0$, then there exists a second vector field $\boldsymbol{a}$ such that

$$\boldsymbol{v} = \operatorname{curl} \boldsymbol{a}. \tag{3.21}$$

Neither ϕ nor $\boldsymbol{a}$ is uniquely determined by $\boldsymbol{v}$. In the first case, if (3.20) holds, then it also holds when ϕ is replaced by $\phi' = \phi + f$, where f is a function of time alone; in the second, if (3.21) holds, then it also holds when $\boldsymbol{a}$ is replaced by

$$\boldsymbol{a}' = \boldsymbol{a} + \operatorname{grad} u,$$

for any scalar function u of position and time.

Aside

It should be kept in mind that the existence statements are local; the proof in the appendix only establishes the existence of ϕ or $\boldsymbol{a}$ in a convex open set. If $\boldsymbol{v}$ is defined on a region U with nontrivial topology, then it may not be possible to find a suitable ϕ or $\boldsymbol{a}$ throughout the whole of U (Exercise 3.5).

Suppose now that we are given fields $\boldsymbol{E}$ and $\boldsymbol{B}$ satisfying Maxwell's equations (2.22)–(2.25) with sources represented by the charge density ρ and the current density $\boldsymbol{J}$. Since $\operatorname{div} \boldsymbol{B} = 0$, there exists a time-dependent vector field $\boldsymbol{A}(t, x, y, z)$ such that

$$\boldsymbol{B} = \operatorname{curl} \boldsymbol{A}.$$

If we substitute $\boldsymbol{B} = \operatorname{curl} \boldsymbol{A}$ into (2.25) and interchange curl with the time derivative, then we obtain

$$\operatorname{curl}\left(\boldsymbol{E} + \frac{\partial \boldsymbol{A}}{\partial t}\right) = 0.$$

It follows that there exists a scalar $\phi(t, x, y, z)$ such that

$$\boldsymbol{E} = -\operatorname{grad}\phi - \frac{\partial \boldsymbol{A}}{\partial t}. \tag{3.22}$$

Definition 3.3

A vector field $\boldsymbol{A}$ such that $\boldsymbol{B} = \operatorname{curl}\boldsymbol{A}$ is a *magnetic vector potential.* A function ϕ such that Equation (3.22) holds is an *electric scalar potential.*

Conversely, given scalar and vector functions ϕ and $\boldsymbol{A}$ of t, x, y, z, we can define $\boldsymbol{B}$ and $\boldsymbol{E}$, by

$$\boldsymbol{B} = \operatorname{curl}\boldsymbol{A}, \qquad \boldsymbol{E} = -\operatorname{grad}\phi - \frac{\partial \boldsymbol{A}}{\partial t}. \tag{3.23}$$

Then two of Maxwell's equations hold automatically, since

$$\operatorname{div}\boldsymbol{B} = 0, \qquad \operatorname{curl}\boldsymbol{E} + \frac{\partial \boldsymbol{B}}{\partial t} = 0.$$

The remaining pair translate into conditions on $\boldsymbol{A}$ and ϕ. Equation (2.22) becomes

$$\operatorname{div}\boldsymbol{E} = -\nabla^2\phi - \frac{\partial}{\partial t}(\operatorname{div}\boldsymbol{A}) = \frac{\rho}{\epsilon_0}$$

and Equation (2.24) becomes

$$\begin{aligned} \operatorname{curl}\boldsymbol{B} - \frac{1}{c^2}\frac{\partial \boldsymbol{E}}{\partial t} &= -\nabla^2\boldsymbol{A} + \operatorname{grad}\operatorname{div}\boldsymbol{A} + \frac{1}{c^2}\frac{\partial}{\partial t}\left(\operatorname{grad}\phi + \frac{\partial \boldsymbol{A}}{\partial t}\right) \\ &= \mu_0\boldsymbol{J}, \end{aligned}$$

where we have used the expansion (B.7) of curl curl. If we put

$$\alpha = \frac{1}{c^2}\frac{\partial \phi}{\partial t} + \operatorname{div}(\boldsymbol{A}),$$

then we can rewrite the equations for $\boldsymbol{A}$ and ϕ more simply as

$$\begin{aligned} \Box\phi - \frac{\partial \alpha}{\partial t} &= \frac{\rho}{\epsilon_0} \\ \Box\boldsymbol{A} + \operatorname{grad}\alpha &= \mu_0\boldsymbol{J}. \end{aligned}$$

Here we have four equations (one scalar, one vector) in four unknowns (ϕ and the components of $\boldsymbol{A}$). Any set of solutions $\phi, \boldsymbol{A}$ determines a solution of Maxwell's equations via (3.23).

3.7 Gauge Transformations

Given solutions $\boldsymbol{E}$ and $\boldsymbol{B}$ of Maxwell's equations, what freedom is there in the choice of $\boldsymbol{A}$ and ϕ? First, $\boldsymbol{A}$ is determined by $\operatorname{curl} \boldsymbol{A} = \boldsymbol{B}$ up to the replacement of $\boldsymbol{A}$ by

$$\boldsymbol{A}' = \boldsymbol{A} + \operatorname{grad} u$$

for some function u of position and time. The scalar potential ϕ' corresponding to $\boldsymbol{A}'$ must be chosen so that

$$\begin{aligned} -\operatorname{grad} \phi' &= \boldsymbol{E} + \frac{\partial \boldsymbol{A}'}{\partial t} \\ &= \boldsymbol{E} + \frac{\partial \boldsymbol{A}}{\partial t} + \operatorname{grad}\left(\frac{\partial u}{\partial t}\right) \\ &= -\operatorname{grad}\left(\phi - \frac{\partial u}{\partial t}\right). \end{aligned}$$

That is, $\phi' = \phi - \partial u/\partial t + f(t)$ where f is a function of t alone. We can absorb f into u by subtracting

$$\int f \, \mathrm{dt}$$

(this does not alter $\boldsymbol{A}'$). So the freedom in the choice of $\boldsymbol{A}$ and ϕ is to make the transformation

$$\boldsymbol{A} \mapsto \boldsymbol{A}' = \boldsymbol{A} + \operatorname{grad} u, \qquad \phi \mapsto \phi' = \phi - \frac{\partial u}{\partial t} \tag{3.24}$$

for any $u = u(t, x, y, z)$.

Definition 3.4

The transformation (3.24) is called a *gauge transformation.*

Under (3.24),

$$\alpha \mapsto \alpha' = \frac{1}{c^2}\frac{\partial \phi'}{\partial t} + \operatorname{div}(\boldsymbol{A}') = \alpha - \Box u.$$

It is possible to show, under certain very mild conditions on α, that the inhomogeneous wave equation

$$\Box u = \alpha \tag{3.25}$$

has a solution $u = u(t, x, y, z)$. If we choose u so that (3.25) holds, then the transformed potentials $\boldsymbol{A}'$ and ϕ' satisfy

$$\operatorname{div}(\boldsymbol{A}') + \frac{1}{c^2}\frac{\partial \phi'}{\partial t} = 0.$$

This is the *Lorenz gauge condition*, named after L. Lorenz (not the H. A. Lorentz of the 'Lorentz contraction').

If we impose the Lorenz condition, then the only remaining freedom in the choice of $\boldsymbol{A}$ and ϕ is to make gauge transformations (3.24) in which u is a solution of the wave equation $\Box u = 0$. Under the Lorenz condition, Maxwell's equations take the form

$$\Box\phi = \rho/\epsilon_0, \qquad \Box\boldsymbol{A} = \mu_0\boldsymbol{J}. \tag{3.26}$$

Consistency with the Lorenz condition follows from the continuity equation on ϕ and $\boldsymbol{J}$.

In the absence of sources, therefore, Maxwell's equations for the potential in the Lorenz gauge reduce to

$$\Box\phi = 0, \qquad \Box\boldsymbol{A} = 0, \tag{3.27}$$

together with the constraint

$$\operatorname{div}\boldsymbol{A} + \frac{1}{c^2}\frac{\partial\phi}{\partial t} = 0.$$

We can, for example, choose three arbitrary solutions of the scalar wave equation for the components of the vector potential, and then define ϕ by

$$\phi = c^2\int \operatorname{div}\boldsymbol{A}\,\mathrm{d}t.$$

Whatever choice we make, we shall get a solution of Maxwell's equations, and every solution of Maxwell's equations (without sources) will arise from some such choice.

3.8 Photons

Maxwell's theory predicts the existence of electromagnetic waves. These are observed, for example, as

radio waves	$\omega < 10^9$
light	$2.5 \times 10^{15} < \omega < 5 \times 10^{15}$
X-rays	$\omega > 10^{18}$

in units of radians per second. In everyday life, we come across waves with ω between 10^{18}, in the X-ray departments of hospitals, and 10^6, Radio 4 on long-wave. The wave velocity c is the velocity of light.

We shall use 'light' to cover electromagnetic radiation of any frequency. Light will play an important part in special relativity, but it is not always convenient to think of it in terms of monochromatic plane waves, which are evenly spread out over all of space. By superimposing plane waves with slightly different frequencies—all close to some ω—we can form a wave packet, which looks like a small localized 'lump' of light. We shall call such a wave packet a *photon*. Photons move with velocity c and have more or less definite frequency ω. Unfortunately, the more we try to localize the photon in space the less precise we can be about the value of ω since we must include in the packet plane waves with a wider and wider spread around ω. This is not a problem provided that the length scales in the problem are large compared with the wavelength $2\pi c/\omega$.

In classical electromagnetism, the idea of a 'photon' is no more than a convenient device to localize problems of light propagation. But in the quantum theory of electrodynamics, light, like other forms of matter, has a dual nature. It has both wave and particle properties, and the 'photon' picture becomes essential.

In quantum theory, each photon has an energy related to its frequency by $E = \hbar\omega$ ($\hbar = 1.05 \times 10^{-34}$ Js).

3.9 Relativity and the Propagation of Light

The principles of electromagnetic theory from which we derived Maxwell's equations are consistent with relativity provided that we look only at the interactions of charged particles moving at speeds much less than that of light. When we look at the propagation of light itself, however, conflict emerges in a very conspicuous way.

It follows from Maxwell's equations that photons—wave packets of light—move with velocity c, whatever direction they travel. In Galilean relativity, however, velocity is additive. If we have two frames of reference in relative motion, then the velocity of a particle relative to the first is the vector sum of its velocity relative to the second and the velocity of the second frame relative to the first. This should apply equally to photons, which would imply that Maxwell's equations could not hold in all frames, since if they did, then the speed of a photon would be independent of the motion of the frame.

There would of course be no inconsistency if we followed Maxwell and adopted the *ether hypothesis*, according to which the equations hold only in the rest frame of the ether; in other frames the velocity of photons depends on direction, just as the velocity of sound depends on direction in a frame which is not at rest relative to the air. The problem is that this resolution is inconsistent with experiment and with common sense, for the following reasons.

- Mechanical forces—for example, between two colliding bodies—are electromagnetic in origin. It is unreasonable that the principle of relativity should apply to mechanics, but not to the underlying electromagnetic processes.
- It is strange that an inconsistency between Galilean relativity and electrodynamics should emerge for rapidly changing fields and rapidly moving frames of reference, when many simple electromagnetic phenomena show a striking indifference to motion through the ether. Einstein gave the example of the current induced in a conductor by a moving magnet; it is the same whether the conductor is at rest and the magnet moving or the magnet at rest and the conductor moving.
- In 1887, Michelson and Morley attempted to detect the direction dependence of the velocity of light in a frame moving relative to the ether, and failed to observe it.

3.10 The Michelson–Morley Experiment

To understand this classic experiment, and the significance of its negative result, let us look in more detail at the 'direction dependence' of the velocity of light predicted by the ether hypothesis.

Suppose that Maxwell's equations hold in a frame R. Then the components (u_1, u_2, u_3) of the velocity of a photon relative to R must satisfy

$$u_1^2 + u_2^2 + u_3^2 = c^2 ,$$

since the speed is c, in every direction. Suppose that we look at the motion from a second frame R', which is moving relative to R in the x-direction. If the coordinates in R are related to those in R' by the Galilean transformation,

$$t = t', \quad x = x' + vt', \quad y = y', \quad z = z', \tag{3.28}$$

then the velocity of the photon relative to R' has components

$$(u_1', u_2', u_3') = (u_1 - v, u_2, u_3) . \tag{3.29}$$

This gives the relationship between the velocities of the photon relative to the frames.

- For a photon moving along the x'-axis of the frame R', for which $u_2' = u_3' = 0$, we have $u_1 = \pm c$ and $u_1' = \pm c - v$. Thus a photon moving in the positive x' direction has speed $c - v$ relative to R', and one moving in the negative x' direction has speed $c + v$ relative to R'.

– For a photon moving along the y'-axis of the frame R', we have $u'_1 = u'_2 = 0$, and therefore $u_1 = v$ and

$$u_3 = u'_3 = \pm\sqrt{c^2 - v^2}\,.$$

Equally for motion in any direction orthogonal to the x'-axis, the speed relative to R' is $\sqrt{c^2 - v^2}$. We can approximate this by using

$$\sqrt{c^2 - v^2} = c\left(1 - \frac{v^2}{2c^2} + O(v^4/c^4)\right)\,.$$

In R', therefore, the velocity of photons is direction-dependent and so Maxwell's equations cannot hold in both R and R'. We shall see later that the false step is the use of the Galilean transformation (3.28), rather than the Lorentz transformation.

The Michelson–Morley experiment was intended to detect the direction-dependence of light velocity due to the assumed motion of the earth relative to ether. Simplifying a little, the apparatus consisted of a slit source S of light of wavelength λ, two plane mirrors A and B, a half-silvered mirror C and a telescope T (Figure 3.2). The light from S was split into two beams at C, which let through some of the light, but reflected the rest. One beam arrived at T after reflection at C and A, the other after reflection at B and C. The two beams interfered at T, producing a pattern of fringes which could be seen in the telescope.

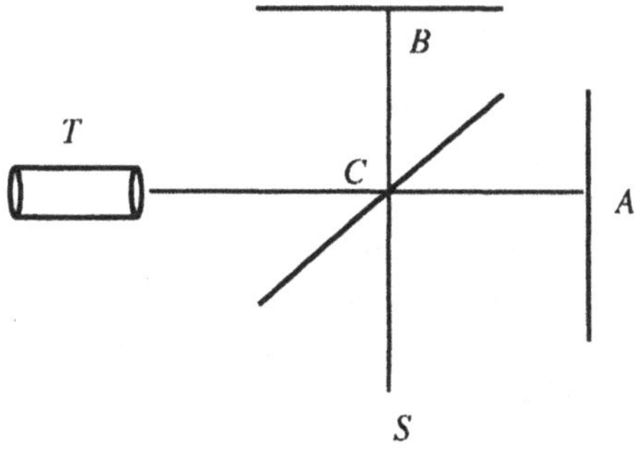

Figure 3.2 The Michelson–Morley experiment

The whole apparatus was attached to a stone disc, which could be rotated. If the earth had been moving through the ether, then the velocity of light in the frame of the earth would have been direction-dependent and the time taken for light to travel from S to T along the two paths would have changed as the stone was rotated.

In the notation above, the time taken for a photon to make the round trip from the origin to the point $x' = D$ on the x'-axis and back is

$$\frac{D}{c+v} + \frac{D}{c-v} = \frac{2D}{c}\left(1 + \frac{v^2}{c^2}\right) + O(v^4/c^4).$$

For the round trip from the origin to $y' = D$ on the y'-axis and back, it is

$$\frac{2D}{\sqrt{c^2 - v^2}} = \frac{2D}{c}\left(1 + \frac{v^2}{2c^2}\right) + O(v^4/c^4).$$

If in the original configuration of the apparatus, the earth had velocity v relative to the ether in the direction CA, then the time taken for light to travel the route CAC would have decreased by

$$\frac{v^2 CA}{c^3} + O(v^4/c^4)$$

when the stone was then rotated through 90°; similarly, the time for route CBC would have increased by

$$\frac{v^2 CB}{c^3} + O(v^4/c^4).$$

Thus, to the second order in v/c, the rotation should have produced an effect equivalent to increasing the difference in the total distances travelled by the two routes by

$$n = \frac{v^2}{c^2 \lambda}(CA + CB) \tag{3.30}$$

wavelengths. There should have been a corresponding shift in the interference pattern.

No fringe shift was observed either in the original experiment or in the many subsequent repetitions and refinements, suggesting either that the earth dragged the ether along with it in its orbit around the sun—a possibility that was rapidly ruled out—or that the assumptions underlying the derivation of Equation (3.30) were incorrect.

EXERCISES

3.3. Suppose that $\rho(\boldsymbol{r})$ is continuously differentiable, and vanishes outside and on the boundary of a bounded region V. Define $\phi(\boldsymbol{r})$ by

$$\phi(\boldsymbol{r}) = \int_{\boldsymbol{r}' \in V} \frac{k\rho(\boldsymbol{r}')}{|\boldsymbol{r} - \boldsymbol{r}'|} \, \mathrm{d}V'.$$

Show that

$$\operatorname{grad}\phi = \int_{r'\in V} \frac{k\,\operatorname{grad}'\rho(\boldsymbol{r}')}{|\boldsymbol{r}-\boldsymbol{r}'|}\,\mathrm{d}V'$$

and hence that $\nabla^2\phi = -4\pi k\rho$ by filling in the following steps. [A prime on a vector operator means that it is applied with $\boldsymbol{r}'$ varying and $\boldsymbol{r}$ fixed.]

(i) Show that

$$\operatorname{grad}|\boldsymbol{r}-\boldsymbol{r}'| = -\operatorname{grad}'|\boldsymbol{r}-\boldsymbol{r}'|\,.$$

(ii) Show that

$$\nabla^2\left(\frac{1}{|\boldsymbol{r}-\boldsymbol{r}'|}\right) = \nabla'^2\left(\frac{1}{|\boldsymbol{r}-\boldsymbol{r}'|}\right) = 0\,,$$

(iii) Show that

$$\nabla^2\phi = -\lim_{\epsilon\to 0}\int_{S_\epsilon} k\rho\,\operatorname{grad}'\left(\frac{1}{|\boldsymbol{r}-\boldsymbol{r}'|}\right).\mathrm{d}\boldsymbol{S}'\,,$$

where S_ϵ is the boundary of the ball B_ϵ, radius ϵ and centre $\boldsymbol{r}$.

3.4. Let $\boldsymbol{a}$, $\boldsymbol{b}$ be constant complex 3-vectors, and let $\Omega = \omega(t - \boldsymbol{e}.\boldsymbol{r}/c)$, where ω and $\boldsymbol{e}$ are real and constant. Under what conditions do the *complex* vectors

$$\boldsymbol{E} = \boldsymbol{a}\,\mathrm{e}^{-\mathrm{i}\Omega}, \quad \boldsymbol{B} = c^{-1}\boldsymbol{b}\,\mathrm{e}^{-\mathrm{i}\Omega}$$

satisfy Maxwell's equations without sources? Show that every monochromatic plane wave can be expressed as the real part of such a complex solution. What are the conditions on $\boldsymbol{a}$ for the plane wave to be (i) linearly polarized and (ii) circularly polarized?

3.5. Show that if $\boldsymbol{B} = \boldsymbol{r}/r^3$ on $\mathbb{R}^3 \setminus \{0\}$, then $\operatorname{div}\boldsymbol{B} = 0$. By considering

$$\int \boldsymbol{B}.\mathrm{d}\boldsymbol{S}$$

over the unit sphere, show that there does not exist a vector field $\boldsymbol{A}$ on $\mathbb{R}^3 \setminus \{0\}$ such that $\boldsymbol{B} = \operatorname{curl}\boldsymbol{A}$.

Why is this not a counterexample to Proposition B.1, Corollary B.3. Which steps taken in the proof are not legitimate in this case?

3.6. Show that if F is a scalar function of $\boldsymbol{r}$ and $\boldsymbol{\alpha}$ is a constant vector, then

$$\nabla^2(F\boldsymbol{\alpha}\wedge\boldsymbol{r}) = \nabla^2(F)\boldsymbol{\alpha}\wedge\boldsymbol{r} + 2\boldsymbol{\alpha}\wedge\operatorname{grad} F.$$

Let $\boldsymbol{b}$ be a vector field depending on $\boldsymbol{r} = (x, y, z)$, but not t, and let ω be a constant. Put

$$\boldsymbol{e} = c^2\omega^{-1}\operatorname{curl}\boldsymbol{b}.$$

Show that $\boldsymbol{B} = \boldsymbol{b}\cos(\omega t)$, $\boldsymbol{E} = \boldsymbol{e}\sin(\omega t)$ satisfy the source-free Maxwell equations if and only if

$$\operatorname{div}\boldsymbol{b} = 0, \quad \text{and} \quad \nabla^2\boldsymbol{b} + k^2\boldsymbol{b} = 0,$$

where $k = \omega/c$. Show that these are satisfied by $\boldsymbol{b} = r^{-2}g(kr)\,\boldsymbol{\alpha}\wedge\boldsymbol{r}$, where $g(x)$ is a function of a single variable, $r = |\boldsymbol{r}|$, and $\boldsymbol{\alpha}$ is a constant vector, if and only if

$$\frac{\mathrm{d}^2g}{\mathrm{d}x^2} + g - \frac{2g}{x^2} = 0.$$

Find a solution that is well behaved at the origin.

4
Einstein's Special Theory of Relativity

4.1 Lorentz's Contraction

Maxwell's electrodynamic theory is an accurate description of the interactions of charged particles and fields, but it is inconsistent with Galilean relativity. It seems to predict that in a moving frame of reference, light travels at different speeds in different directions; and it seems to require an 'ether'—an all pervasive medium that determines an absolute standard of rest. But the ether defies detection; the Michelson–Morley experiment, and all subsequent experiments designed to measure motion relative to the ether, gave negative results.

H. A. Lorentz suggested a way out: he argued that a rigid body—in the case of the Michelson–Morley experiment, the stone—should contract by a factor

$$\sqrt{1 - v^2/c^2}$$

along the direction of its motion through the ether as the result of a supposed effect of the motion on the electromagnetic forces between the particles making up the body. The contraction would reduce the fringe shift to the extent that it would be unobservable. The idea was extended by Lorentz after the failure of other ether detection experiments and was put by Poincaré into the form of a fundamental principle that no experiment could detect motion through the ether; any effect that might be detectable would be exactly cancelled by an equal compensating effect. There was an analogy with Newton's third law: to every action there is an equal and opposite reaction.

According to Poincaré's point of view, the ether exists, but cannot be detected. This is not a good starting point for a physical theory. It also raises a still more awkward problem; it raises doubts about the physical meaning of distance measurements. A moving measuring rod contracts by a factor

$$\sqrt{1 - v^2/c^2}\,.$$

Thus only measurements made in a frame at rest relative to the ether are valid. But if the ether is undetectable, how is distance to be measured?

Einstein saw that the problem lay not with Maxwell's equations nor with the principle of relativity, but with the uncritical acceptance of intuitive ideas about the measurement of space and time. He saw that one cannot resolve the issues raised by the conflict with relativity without first explaining what is meant by 'length' in terms of the process of measurement of distance; that is, 'distance' must be given an *operational definition*—a definition in terms of the operations required to measure it.

Before tackling distance, however, it is necessary first to give an operational definition of 'simultaneity', because even in the classical view of space and time, distance is only defined between simultaneous events. (See Chapter 1: what is the distance between Oxford at 2 pm and Cambridge at 3 pm, measured in a frame fixed relative to the sun?)

Einstein began his analysis of electrodynamics in moving frames by giving operational definitions of 'distance' and 'simultaneity' which were consistent with the principle of relativity and with the validity of Maxwell's equations in all uniformly moving frames of reference. It followed that the time and distance coordinates in different frames could not be related by Galilean transformations.

We shall use slightly different, but essentially equivalent, operational definitions due to E. A. Milne [5]. In this chapter, we shall look at how they lead to the replacement of Galilean transformations by *Lorentz transformations* in one space dimension—that is, in the case of motion along a line. The argument is a very elegant one, due to H. Bondi [1].

4.2 Operational Definitions of Distance and Time

In Milne's approach, one takes 'clocks' and 'light signals' as fundamental. Every observer carries a clock with which he can measure the time of events in his immediate vicinity and observers can send out and receive light signals, which are carried by photons (particles of light). The vibrations of a single atom can be used to measure time, so a clock is intrinsically a much simpler object than a measuring rod made up of a very large number of atoms.

Since Maxwell's equations are to hold in every frame, the definitions of distance and simultaneity must be such that the following is true.

Velocity of Photons

The velocity of photons is the same irrespective of the motion of their source or of the observer.

A non-accelerating observer moving along a straight line can use his clock and light signals to assign coordinates t and x to distant events on the line. Suppose that he sends out a light signal at time t_1 (measured on his clock). This is received at an event A on the line and immediately transmitted back to the observer, arriving at time t_2 (again measured on the observer's clock). Which event B at the observer is simultaneous with A? If the velocity of photons is assumed to be constant, then the journeys of the outgoing and returning photons will be reckoned by the observer to have equal duration, and so the observer will take B to be the event at his location which happens at time $\frac{1}{2}(t_1 + t_2)$ and he will assign this value of t to A. This is the *radar definition* of simultaneity. It is illustrated in the space–time diagram Figure 4.1, in which the vertical line is the worldline of the observer and the lines at 45° are the worldlines of the outgoing and returning photons.

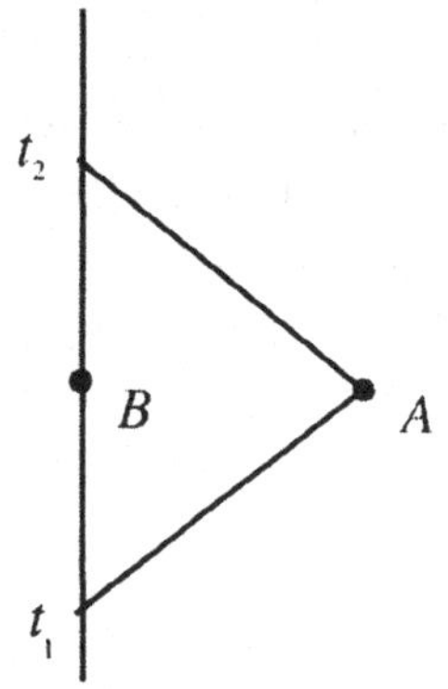

Figure 4.1 Milne's definitions

Definition of Simultaneity and Distance

The observer defines A to be simultaneous with the event B on his world-line that happens at time $\frac{1}{2}(t_1 + t_2)$ and assigns a distance $\frac{1}{2}c(t_2 - t_1)$ to the separation of B from A.

Here c is a constant which is chosen arbitrarily, but is given the same value by all observers. If t is measured in seconds and c is chosen to be 3×10^8, then the unit of distance is called the *metre.*[1] It t is measured in years and $c = 1$, then the unit of distance is the light-year; and so on. By defining distance and simultaneity in this way, a non-accelerating observer can, in principle, set up a coordinate system to label each event by

- its radar distance x from his own location, measured positively in one direction along the line and negatively in the other; and
- the time t at which it happens, according to the radar definition.

These labels are called *inertial coordinates.* We shall always assume that the observer sets $x = 0$ at his own location.

4.3 The Relativity of Simultaneity

The advantage in accepting these definitions is that it follows automatically that the velocity of light is independent of the observer, so the null result of the Michelson–Morley experiment is no longer a problem. The disadvantage is that we must also accept various consequences which are contrary to intuition. In particular, we must accept that simultaneity is relative. Two events which are reckoned to be simultaneous by one observer O may not be simultaneous according to a second observer O' moving relative to O. This is easiest to see from the space–time diagram, Figure 4.2. Here each point represents an event and time increases up the page. The two solid lines are the histories of O and O' and the thin lines are the histories of photons. More prosaically, one can think of the various lines as graphs of time as functions of position.

Two light signals are transmitted from the event C where O' passes O. They are reflected at A_1 and A_2 and both arrive back at O at the event D. The observer O therefore judges the two events A_1 and A_2 to be simultaneous. However, the photon from A_2 reaches O' before that from A_1, so O' assigns an earlier time to A_2 than to A_1; he judges that the two events are not simultaneous.

[1] In fact c is taken to be *exactly* 299,792,458 in SI units.

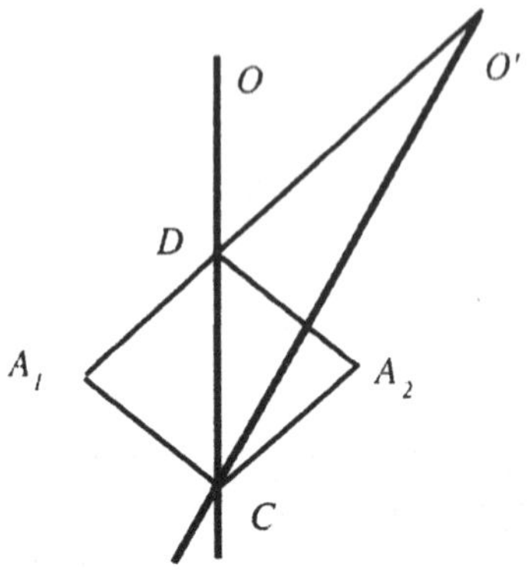

Figure 4.2 The relativity of simultaneity

In the following example, we look at the relativity of simultaneity in a slightly different, but more quantitative way.

Example 4.1

Suppose that O sets up the inertial coordinate system x, t and that O' passes O at $t = 0$. Then the worldline of O' is given by

$$x = ut$$

for some constant u, which O will interpret as the velocity of O'. See Figure 4.3.

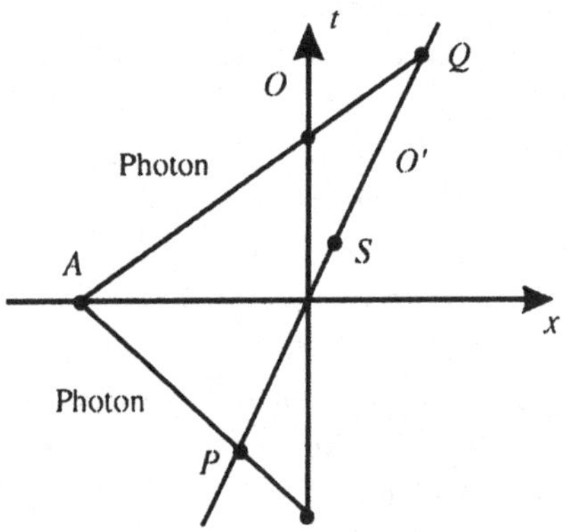

Figure 4.3 Example 4.1

Consider the event A with coordinates

$$t = 0, \qquad x = -D \quad (D > 0).$$

A photon that reaches $x = -D$ at $t = 0$ must have left O at time $t = -D/c$ (measured on the clock carried by O). This photon passes O' at the event P with coordinates x_P, t_P, where

$$x_P = ut_P = -D - ct_P .$$

The first equality holds because P lies on the worldline of O'; the second because it lies on the worldline of a photon travelling with speed c in the negative x-direction and with $x = 0$ at $t = -D/c$. Thus P has coordinates

$$x_P = -\frac{uD}{u+c}, \qquad t_P = -\frac{D}{u+c} .$$

Similarly, a photon emitted at A and travelling with speed c in the positive x-direction reaches O at the event with coordinates $(x,t) = (0, D/c)$ and reaches O' at the event Q with coordinates

$$x_Q = \frac{uD}{c-u}, \qquad t_Q = \frac{D}{c-u} .$$

As one would expect, O thinks that A is simultaneous with the origin $(0,0)$ of the inertial coordinates, since this is the event at his own location that happens at time

$$t = \frac{1}{2}\left(\frac{D}{c} - \frac{D}{c}\right) = 0 .$$

On the other hand, O' thinks that A happens simultaneously with the event at *his* location which happens midway between P and Q. That is, he reckons A is simultaneous with the event S with coordinates

$$x_S = \tfrac{1}{2}(x_P + x_Q) = \frac{u^2 D}{c^2 - u^2}, \qquad t_S = \tfrac{1}{2}(t_P + t_Q) = \frac{uD}{c^2 - u^2} .$$

Since $x_S = ut_S$, this event is indeed on the worldline of O'. However, t_S is nonzero unless either D or u vanishes, so S is not simultaneous with A according to O. Our two observers O and O' have different notions of simultaneity.

It is interesting to put in some numbers. If, say, D is a 10 m and u is 10 m/s, then with $c = 3 \times 10^8$, we get

$$t_S \sim 10^{-15}\,\mathrm{s} ;$$

that is, a femtosecond—a millionth of a nanosecond—in contrast to the 'Galilean' value of zero.

Over everyday speeds and distances, the relativity of simultaneity does not have a significant effect. To get something more easily observable, either u must be a substantial fraction of the velocity of light—as is the case, for example, with particles in a modern accelerator—or D must be large. If, for example, D is 10,000,000 light-years and u is still 10 m/s, then

$$t_S \sim \frac{10 \times 10^7}{3 \times 10^8} = \frac{1}{3} \quad \text{years} .$$

So even if the two observers have a relative speed as low as 10 m/s, their notions of simultaneity over inter-galactic distances are significantly different—in this case, by four months over a distance of 10 million light-years.

Of course in neither case is it practical actually to perform the operations involved in the radar method; but in principle they determine the relativistic notions of time and distance.

4.4 Bondi's k-Factor

If two observers set up inertial coordinate systems on space–time by using Milne's radar method to determine the location and time of distant events, then the two systems cannot be related by a Galilean transformation, since the Galilean transformation is incompatible with the assumption that the velocity of light is independent of the observer. How then are they related?

We shall answer this first in a world of one space and one time dimension, where all motion is along a single straight line. So consider two observers O and O' travelling along the line with constant speeds. Suppose that they pass each other at the event E and then move directly away from each other. Suppose also that they both set their clocks to zero at E.

By using the radar definitions, they both set up inertial coordinate systems on two-dimensional space–time: O will label the events on the line by their distance x along the line, and by the time at which they happen, according to his measurements; and likewise O' will label them by x' and t'.

We shall derive the relationship between (x, t) and (x', t') by making two assumptions.

- Both observers reckon that the velocity of light is c.
- Only their relative motion is observable.

We begin by considering a photon emitted by O towards O' at time t (measured on the clock that O carries). Suppose that it is received by O' at time $t' = kt$ (measured on the clock carried by O'). The quantity k is called *Bondi's k-factor*.

Since neither observer is accelerating, k is constant and, as a consequence of the second assumption, *k depends only on the relative velocity of O and O'*. It is in this last innocuous looking statement that we depart from classical ideas.

4.5 Time Dilation

Because k depends only on the relative motion, we have the following.

- A photon sent by O towards O' at time t (measured on the clock carried by O) arrives at O' at time $t' = kt$ (measured on the clock carried by O').
- A photon sent by O' towards O at time t' (measured on the clock carried by O') arrives at O at time $t = kt'$ (measured on the clock carried by O).

Now consider the space–time diagram Figure 4.4. Here a photon sent by O at

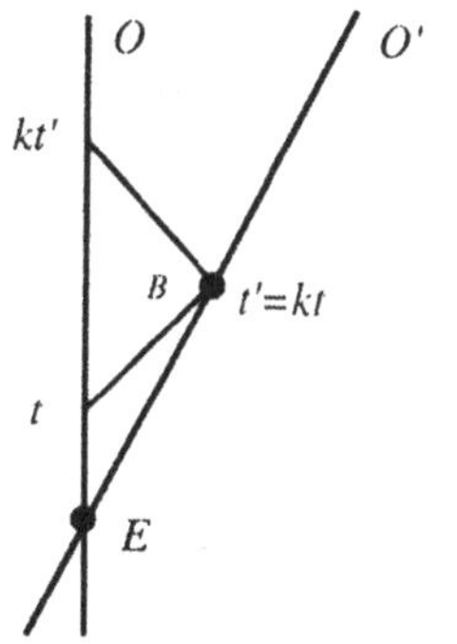

Figure 4.4 The k-factor

time t measured on the clock carried by O arrives at O' at event B, which happens at time kt measured on the clock carried by O'; it is then sent back to O, arriving at time k^2t, as measured on the clock carried by O. Hence O measures the distance from his location to B and the time of B to be

$$d_B = \tfrac{1}{2}c(k^2 - 1)t, \qquad t_B = \tfrac{1}{2}(1 + k^2)t\,.$$

Thus O reckons that the speed of O' is

$$u = \frac{d_B}{t_B} = \frac{c(k^2 - 1)}{k^2 + 1}\,.$$

Solving for k, we have

$$k = \sqrt{\frac{c+u}{c-u}} > 1\,.$$

It follows that

$$\frac{\text{time } E \text{ to } B \text{ measured by } O}{\text{time } E \text{ to } B \text{ measured by } O'} = \frac{t_B}{kt} = \frac{(k^2+1)t}{2kt} = \gamma(u)\,,$$

where the *gamma factor* $\gamma(u)$ is defined by

$$\gamma(u) = \frac{1}{\sqrt{1-u^2/c^2}}\,.$$

Example 4.2

An astronaut is travelling directly away from earth with speed $u = c\sqrt{3}/2$, so that

$$\frac{1}{\sqrt{1-u^2/c^2}} = 2\,.$$

So if the astronaut reckons that one hour passes between two events in his spaceship, then an observer on earth reckons that two hours pass.

This is the *time dilation effect*; the time between the two events depends on the observer. It only becomes paradoxical if one insists on talking about 'time' independently of the process of measurement of time. It is also important to realize that the apparent asymmetry between the observer on earth and the astronaut arises because both are measuring the time interval between events on the astronaut's worldline. If we consider two events on earth (say at $t = 0$ and at $t = 1$), then the situation is reversed: O assigns a unit time separation; and the astronaut O' assigns a time separation $\gamma(u) > 1$.

4.6 The Two-dimensional Lorentz Transformation

We shall now consider how the coordinate systems set up by two observers are related. We shall consider first the case of observers moving uniformly on the same straight line, so we have a two-dimensional space–time—events are points on the line at particular times. By using the radar method, each observer can assign inertial coordinates x, t to an event; x is the distance from the observer (with an appropriate sign) and t is the time of the simultaneous event on the observer's own worldline.

How are the inertial coordinate systems x, t and x', t' of the two observers O and O' related? For simplicity, we shall assume that both set their clocks to zero at the event E at which they pass. Then E will be the common origin of the two coordinate systems.

Proposition 4.3

The inertial coordinate systems set up by O and O' are related by

$$\begin{pmatrix} ct \\ x \end{pmatrix} = \gamma(u) \begin{pmatrix} 1 & u/c \\ u/c & 1 \end{pmatrix} \begin{pmatrix} ct' \\ x' \end{pmatrix}, \tag{4.1}$$

where u is the relative velocity and $\gamma(u) = 1/\sqrt{1 - u^2/c^2}$.

This is the (two-dimensional) *Lorentz transformation.*

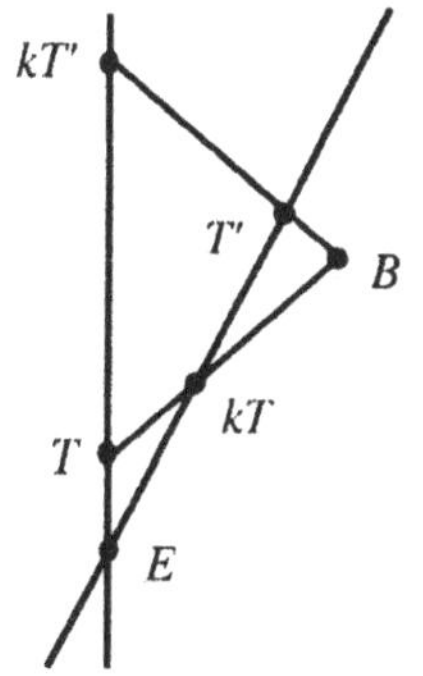

Figure 4.5 The derivation of the coordinate transformation

Proof

Let k denote Bondi's factor. Consider the space–time diagram Figure 4.5. A photon is sent out from O at time T measured on the clock carried by O, passes O' at time kT measured on the clock carried by O', is reflected at the event B, passes O' again at time T' measured on the clock carried by O', and returns to O at time kT' measured on the clock carried by O.

– In the inertial coordinate system of observer O, the coordinates of B are:

$$t = \tfrac{1}{2}(kT' + T), \qquad x = \tfrac{1}{2}c(kT' - T).$$

– In the inertial coordinate system of observer O', the coordinates of B are:

$$t' = \tfrac{1}{2}(T' + kT), \qquad x' = \tfrac{1}{2}c(T' - kT).$$

Hence we have

$$\begin{pmatrix} ct \\ x \end{pmatrix} = \frac{c}{2}\begin{pmatrix} 1 & k \\ -1 & k \end{pmatrix}\begin{pmatrix} T \\ T' \end{pmatrix}$$
$$\begin{pmatrix} ct' \\ x' \end{pmatrix} = \frac{c}{2}\begin{pmatrix} k & 1 \\ -k & 1 \end{pmatrix}\begin{pmatrix} T \\ T' \end{pmatrix},$$

and therefore

$$\begin{aligned}\begin{pmatrix} ct \\ x \end{pmatrix} &= \frac{1}{2k}\begin{pmatrix} 1 & k \\ -1 & k \end{pmatrix}\begin{pmatrix} 1 & -1 \\ k & k \end{pmatrix}\begin{pmatrix} ct' \\ x' \end{pmatrix} \\ &= \frac{1}{2}\begin{pmatrix} k+k^{-1} & k-k^{-1} \\ k-k^{-1} & k+k^{-1} \end{pmatrix}\begin{pmatrix} ct' \\ x' \end{pmatrix}.\end{aligned}$$

But we showed above that $k = \sqrt{(c+u)/(c-u)}$. Therefore

$$k + k^{-1} = \frac{2c}{\sqrt{c^2 - u^2}}, \qquad k - k^{-1} = \frac{2u}{\sqrt{c^2 - u^2}},$$

and the result follows. Some sign choices have been made in relating x and x', which can be positive or negative, to the distances from O or O', which are necessarily positive. □

The relationship between the two coordinate systems is shown in Figure 4.6. If we put $x' = 0$, then $x = ut$. With the sign choices we have made, O' is moving

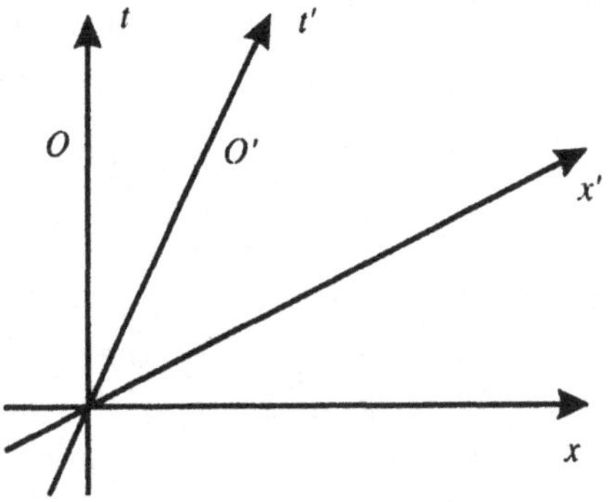

Figure 4.6 The two-dimensional Lorentz transformation

relative to O in the positive x direction with speed u. We also have $t = \gamma(u)t'$ when $x' = 0$, which is the time dilation formula for events on the worldline of O'. The inverse transformation is

$$\begin{pmatrix} ct' \\ x' \end{pmatrix} = \gamma(u) \begin{pmatrix} 1 & -u/c \\ -u/c & 1 \end{pmatrix} \begin{pmatrix} ct \\ x \end{pmatrix},$$

so O is moving relative to O' in the *negative* x' direction with the same speed u.

The transformation reduces to the Galilean transformation when $u \ll c$, since we have

$$\begin{pmatrix} t \\ x \end{pmatrix} = \gamma(u) \begin{pmatrix} 1 & u/c^2 \\ u & 1 \end{pmatrix} \begin{pmatrix} t' \\ x' \end{pmatrix} \to \begin{pmatrix} 1 & 0 \\ u & 1 \end{pmatrix} \begin{pmatrix} t' \\ x' \end{pmatrix}$$

as $c \to \infty$.

4.7 Transformation of Velocity

Consider a non-accelerating particle moving with speed v relative to O in the negative x direction, so that $x = -vt + a$ for some constant a. In classical theory, its speed relative to O' would be $u + v$. In relativity, we have

$$\begin{pmatrix} ct' \\ x' \end{pmatrix} = \gamma(u) \begin{pmatrix} 1 & -u/c \\ -u/c & 1 \end{pmatrix} \begin{pmatrix} ct \\ -vt + a \end{pmatrix},$$

so in the x', t' coordinate system set up by O', the events in the history of the particle are given by

$$\begin{aligned} t' &= \gamma(u) \left[\left(1 + \frac{uv}{c^2}\right) t - \frac{au}{c^2} \right] \\ x' &= \gamma(u) \left[-(u+v)t + a \right]. \end{aligned}$$

Therefore the speed w of the particle relative to O' is

$$w = -\frac{\mathrm{d}x'}{\mathrm{d}t'} = \frac{v + u}{1 + uv/c^2}.$$

This is the *velocity addition formula*. It is not the same as the classical formula $w = v + u$, but reduces to it when $u, v \ll c$. If $v = c$, then $w = c$ irrespective of the value of u, so photons move at speed c relative to all observers. The velocity of the observer is not added to c in a moving frame, as it would be in the classical theory.

Note that if $|u| < c$ and $|v| < c$ then $|w| < c$ since

$$(c-u)(c-v) > 0 \quad \Leftrightarrow \quad u + v < c(1 + uv/c^2) \tag{4.2}$$

$$(c+u)(c+v) > 0 \quad \Leftrightarrow \quad u + v > -c(1 + uv/c^2). \tag{4.3}$$

4.8 The Lorentz Contraction

Consider two observers O and O' whose inertial coordinate systems are related by (4.1). Suppose that a rod lies along the x'-axis between $x' = 0$ and $x' = L$ and is at rest relative to O'. Then according to O', its length is L. What is its length as measured by O?

We must first be clear about what the question means. In the inertial coordinate system of O', the worldlines of the ends of the rod are given by $x' = 0$ and by $x' = L$. In the inertial coordinate system of O, therefore, the two worldlines are given parametrically (with t' as parameter) by

$$\begin{pmatrix} ct \\ x \end{pmatrix} = \gamma(u) \begin{pmatrix} 1 & u/c \\ u/c & 1 \end{pmatrix} \begin{pmatrix} ct' \\ 0 \end{pmatrix} = \gamma(u) \begin{pmatrix} ct' \\ ut' \end{pmatrix} \tag{4.4}$$

$$\begin{pmatrix} ct \\ x \end{pmatrix} = \gamma(u) \begin{pmatrix} 1 & u/c \\ u/c & 1 \end{pmatrix} \begin{pmatrix} ct' \\ L \end{pmatrix} = \gamma(u) \begin{pmatrix} ct' + Lu/c \\ ut' + L \end{pmatrix} . \tag{4.5}$$

The question is: what is the distance measured by O between two events E and B, one on the first worldline, one on the second, which are simultaneous according to O? If we take E to be the event $t = 0$, $x = 0$, then B must be as in (4.5), with t' chosen so that $t = 0$. That is $t' = -Lu/c^2$, which implies that B is the event

$$t = 0, \qquad x = \gamma(u)(-Lu^2/c^2 + L) = L\sqrt{1 - u^2/c^2} . \tag{4.6}$$

So according to O, the rod is shorter by a factor $\sqrt{1 - u^2/c^2}$. This is the same as the factor proposed by Lorentz in the context of the ether hypothesis—hence the term *Lorentz contraction*, or sometimes *FitzGerald–Lorentz contraction* since the Irish mathematician George FitzGerald put forward the same idea a few years before Lorentz.[2] However, the interpretation here is very different. In particular, it is important to remember that when O measures the length, he is assigning a 'distance' to the spatial separation between E and B; but when O' measures the length, he is assigning a 'distance' to the separation between E and A, the event at the other end of the rod simultaneous with E according to his own definition of simultaneity (see Figure 4.7).

4.9 Composition of Lorentz Transformations

We can also derive the *velocity addition formula* from another point of view. Suppose that O, O', O'' are observers moving along the line at constant speed,

[2] George FitzGerald was a pupil of George Boole's sister; the separation of mathematical ideas is sometimes less than one imagines.

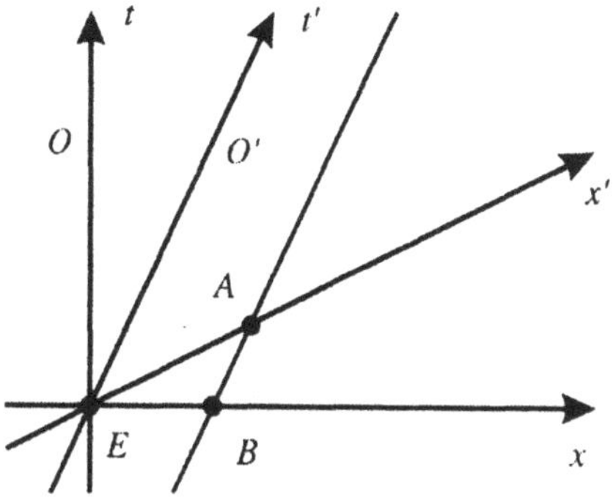

Figure 4.7 The Lorentz contraction

and that they meet at the event E. Then their respective inertial coordinate systems are related by

$$\begin{pmatrix} ct \\ x \end{pmatrix} = \gamma(u)\begin{pmatrix} 1 & u/c \\ u/c & 1 \end{pmatrix}\begin{pmatrix} ct' \\ x' \end{pmatrix} \tag{4.7}$$

$$\begin{pmatrix} ct'' \\ x'' \end{pmatrix} = \gamma(v)\begin{pmatrix} 1 & v/c \\ v/c & 1 \end{pmatrix}\begin{pmatrix} ct \\ x \end{pmatrix} \tag{4.8}$$

$$\begin{pmatrix} ct'' \\ x'' \end{pmatrix} = \gamma(w)\begin{pmatrix} 1 & w/c \\ w/c & 1 \end{pmatrix}\begin{pmatrix} ct' \\ x' \end{pmatrix}, \tag{4.9}$$

where u is the velocity of O' relative to O, v is the velocity of O relative to O'', and w is the velocity of O' relative to O''. So we must have

$$\gamma(w)\begin{pmatrix} 1 & w/c \\ w/c & 1 \end{pmatrix} = \gamma(u)\gamma(v)\begin{pmatrix} 1 & v/c \\ v/c & 1 \end{pmatrix}\begin{pmatrix} 1 & u/c \\ u/c & 1 \end{pmatrix},$$

and therefore

$$\gamma(w) = \gamma(u)\gamma(v)\left(1 + \frac{uv}{c^2}\right). \tag{4.10}$$

However,

$$\frac{w^2}{c^2} = \frac{\gamma(w)^2 - 1}{\gamma(w)^2}.$$

After substituting from (4.10), and doing a little algebra, we find that

$$w^2 = \frac{(u+v)^2}{(1 + uv/c^2)^2},$$

which again gives the velocity addition formula.

4.10 Rapidity

The Lorentz transformation and velocity addition formula take on a more familiar look if we put $\phi(u) = \log k = \tanh^{-1}(u/c)$. Then

$$\begin{pmatrix} ct \\ x \end{pmatrix} = \begin{pmatrix} \cosh\phi & \sinh\phi \\ \sinh\phi & \cosh\phi \end{pmatrix} \begin{pmatrix} ct' \\ x' \end{pmatrix},$$

so a Lorentz transformation is a *hyperbolic rotation*. The quantity ϕ is called the *rapidity* or *pseudo-velocity* of the transformation. It is analogous to the angle of a rotation in the plane.

In terms of rapidity, the velocity addition formula takes the more suggestive form

$$\phi(w) = \phi(u) + \phi(v).$$

4.11 *The Lorentz and Poincaré Groups

The Lorentz transformations in two-dimensional space–time form a group, called the *Lorentz group*; the inverse of a Lorentz transformation and the product of two Lorentz transformations are again Lorentz transformations. Rapidity determines an isomorphism with the additive group of real numbers.

Strictly, this group should be called the *proper orthochronous Lorentz group*: the full group is obtained by composing Lorentz transformations with

- the *space reflection*

$$\begin{pmatrix} ct \\ x \end{pmatrix} \mapsto \begin{pmatrix} ct' \\ x' \end{pmatrix} = \begin{pmatrix} ct \\ -x \end{pmatrix},$$

  which reverse the orientation of the line; and

- the *time reversal*

$$\begin{pmatrix} ct \\ x \end{pmatrix} \mapsto \begin{pmatrix} ct' \\ x' \end{pmatrix} = \begin{pmatrix} -ct \\ x \end{pmatrix}.$$

This gives a *Lie group* with four connected components, each homeomorphic to the real line. Which component a transformation lies in is determined by whether it reverses time, or spatial orientation, or both, or neither.

The (proper, orthochronous) Lorentz group is analogous to the group SO(2) of rotations of the Euclidean plane, which is the group of 2×2 matrices of the form

$$\begin{pmatrix} \cos\theta & \sin\theta \\ -\sin\theta & \cos\theta \end{pmatrix}.$$

This is similarly one component of a larger group, O(2), generated by the rotations and the reflection in the x-axis,

$$\begin{pmatrix} 1 & 0 \\ 0 & -1 \end{pmatrix} .$$

Two important differences are, first, that in Euclidean space the full group has only two components, whereas in space–time it has four; and, second, the rotation group in Euclidean space is compact, while the Lorentz group is not.

The group O(2) can be extended to the *Euclidean group* by adding translations of the plane. The result is the *isometry group* of the Euclidean plane, which is generated by rotations, reflections, and translations. The full Lorentz group can similarly be extended, to give the *Poincaré group*, which is generated by (proper, orthochronous) Lorentz transformations, space reflections, time reversals, and translations. In this book, all Lorentz transformations will be proper and orthochronous, unless we explicitly allow otherwise.

EXERCISES

4.1. Two inertial coordinate systems x, t and x', t' (in one space dimension) are related by the Lorentz transformation

$$\begin{pmatrix} ct \\ x \end{pmatrix} = \gamma(u) \begin{pmatrix} 1 & u/c \\ u/c & 1 \end{pmatrix} \begin{pmatrix} ct' \\ x' \end{pmatrix}, \qquad \gamma(u) = 1/\sqrt{1 - u^2/c^2} .$$

(i) Show that if ψ is a function of x and t, then

$$\frac{1}{c^2}\frac{\partial^2 \psi}{\partial t^2} - \frac{\partial^2 \psi}{\partial x^2} = \frac{1}{c^2}\frac{\partial^2 \psi}{\partial t'^2} - \frac{\partial^2 \psi}{\partial x'^2} .$$

Show that a harmonic solution $\psi = A\cos(\Omega)$, $\Omega = \omega(ct - x)/c$, of the wave equation also has the form of a harmonic wave when expressed in terms of the coordinates x', t'. What is its frequency in the new coordinates?

(ii) Show that if $u \ll c$, then the event with coordinates $t' = 0$, $x' = -D$ in the second system has coordinates $t = -uD/c^2$, $x = -D$ in the first (to the first order in u/c).

(iii) A man passes a woman in the street at a relative speed of about 5 mph. The woman is walking directly away from the Andromeda galaxy and the man is walking directly towards it. As they pass, the woman says 'At this moment a battle fleet is setting out from Andromeda to destroy our galaxy.' How long ago did the battle fleet leave according to the man's reckoning? (Andromeda is about 2,200,000 light-years away.)

4.2. Show that a linear transformation

$$\begin{pmatrix} ct \\ x \end{pmatrix} = L \begin{pmatrix} ct' \\ x' \end{pmatrix}$$

is a Lorentz transformation if and only if (i) the top left entry in L is positive, (ii) $\det L > 0$, and

$$\text{(iii)} \qquad L^{\mathrm{t}} \begin{pmatrix} 1 & 0 \\ 0 & -1 \end{pmatrix} L = \begin{pmatrix} 1 & 0 \\ 0 & -1 \end{pmatrix}.$$

Show that when these conditions are satisfied, the rapidity ϕ of the transformation is given by $\operatorname{tr} L = 2\cosh\phi$.

4.3. †An athlete carrying a 20 ft pole PQ runs with speed $\sqrt{3}c/2$ in the direction PQ into a 10 ft room. Show that, as measured in the frame of the room, the pole is 10 ft long.

As the end Q reaches the far wall, a man standing by the door closes it. Explain with the aid of a space–time diagram how this is possible when in the athlete's coordinate system, the pole has length 20 ft but the room is only 5 ft long.

If the athlete holds the end P, is he (i) outside the room, (ii) inside the room, or (iii) at the door when he first feels the shock of the end Q striking the wall?

4.4. †Two light sources are at rest and at distance D apart on the x-axis of some inertial frame. They emit photons simultaneously in that frame in the positive x direction. Show that in a frame in which the sources have velocity u along the x-axis, the photons are separated by a constant distance

$$D\sqrt{\frac{c-u}{c+u}}.$$

5 Lorentz Transformations in Four Dimensions

5.1 Coordinates in Four Dimensions

An observer travelling in a straight line at constant speed can determine the coordinates t, x of events that happen along the line by the radar method: x is the distance of an event from the observer and t is the time shown on the observer's clock at the simultaneous event at the observer's location. It is built into the definition of x and t that light travels at constant speed; but it is a consequence of it that a second observer moving along the line at a different speed will have a different idea of simultaneity. The coordinate systems of the two observers are related not by the classical Galilean transformation, but by the Lorentz transformation.

We now turn to consider more general relative motion; our observers will still be inertial, and so still moving in straight lines with constant velocity, but they need no longer move along the same straight line.

In order to define the coordinates of events that can happen anywhere in space, an observer needs, in addition to a clock, a device to measure the direction from which light signals arrive. He can then assign polar coordinates to an event: the distance r from his own location and the time of the event are defined by the radar method; and the two polar angular coordinates θ and ϕ are given by the direction of the returning light signal. He recovers the Cartesian coordinates x, y, z of the event from the standard transformation:

$$x = r\sin\theta\cos\phi, \qquad y = r\sin\theta\sin\phi, \qquad z = r\cos\theta\,.$$

The result is an *inertial frame of reference* or *inertial coordinate system* t, x, y, z on space–time, in which the observer's own worldline is given by

$$x = y = z = 0\,.$$

This is not intended to be a practical procedure—it is an *operational definition* that gives a method 'in principle' for setting up coordinates on space–time. Implicit is the assumption that the observer knows how to compare the directions from which light signals arrive at different times; in other words that it makes sense to say that the angle-measuring device is carried *without rotation*. An 'inertial' observer is an observer who is neither accelerating nor rotating. Again, this is a notion that must be re-examined more critically in the context of gravitation, but we shall content ourselves with the idea that both acceleration and rotation can be *felt*, and so there is an absolute meaning to the requirement that they should be absent.

5.2 Four-dimensional Coordinate Transformations

In order to derive the relationship between the coordinate systems t, x, y, z and t', x', y', z' set up by two inertial observers O and O', we have to make some assumptions. It is an interesting challenge to try to reduce these to some minimal self-evident set, but not as easy as in the case of motion along a line, where the Lorentz transformation law was seen to emerge very directly from the properties of Bondi's k-factor. In drawing up our list, we shall not attempt to refine the assumptions down to some basic set of independent laws. Rather, we shall simply make explicit the properties of the Galilean transformation that are assumed to be retained, and, as in the case of linear motion, the properties of light propagation that are built into the kinematical foundations of Einstein's theory. In this spirit, we list our assumptions as follows.

– The transformation is affine linear. That is, it is of the form

$$\begin{pmatrix} ct \\ x \\ y \\ z \end{pmatrix} = L \begin{pmatrix} ct' \\ x' \\ y' \\ z' \end{pmatrix} + C\,, \tag{5.1}$$

where L is a nonsingular 4×4 matrix and C is a column vector.

- Photons travel in straight lines with velocity c relative to any inertial coordinate system. That is, photon worldlines are of the form

$$x = u_1 t + a_1, \quad y = u_2 t + a_2, \quad z = u_3 t + a_3 ,$$

where u_1, u_2, u_3, a_1, a_2, a_3 are constants and $u_1^2 + u_2^2 + u_3^2 = c^2$.

- No physical effect is transmitted faster than light.
- The principle of relativity applies to all physical phenomena—only the relative motion of non-accelerating observers can be detected by physical experiments.

The first assumption is the one property of Galilean transformations that we do not want to throw away. It is equivalent to the assertion that if Newton's first law holds in one coordinate system, then it also holds in the other; in both, the worldlines of free particles—particles not acted on by any force—are straight lines in space–time, given by linear equations. The second incorporates the assumption that the velocity of light should be independent of the observer. It must hold if light propagates by Maxwell's equations in an inertial coordinate system. The third assumption is needed for consistency, as we shall see. The restriction to non-accelerating observers in the fourth is the meaning of the 'special' in 'special relativity'.

We denote the top left entry in L by γ: this is the *time dilation factor* for the motion of O' relative to O; along the worldline of O', which is given by $x' = y' = z' = 0$, we have

$$t = \gamma t' + \text{constant}.$$

So γ relates the time measurements of events on the worldline of O' in the two coordinate systems. Similarly, if γ' is the top left entry in L^{-1}, then along the worldline of O ($x = y = z = 0$) we have

$$t' = \gamma' t + \text{constant}.$$

So γ' is the time dilation factor for the motion of O relative to O'. It follows from the fourth assumption, the relativity assumption, that the time dilation factor depends only on the relative motion of the two observers, and hence that $\gamma = \gamma'$.

It follows from the second assumption that the worldlines of photons through an event A form a cone in space–time. This is called the *light-cone* of the event. If A is the event $t = x = y = z = 0$, then the light-cone of A has equation

$$c^2t^2 - x^2 - y^2 - z^2 = 0 , \tag{5.2}$$

which is the condition that the the time t elapsed from A at light speed should be related to the distance $D = \sqrt{x^2 + y^2 + z^2}$ from A by $D = ct$. The light-cone consists of the event A itself, together with the *future light-cone*, made

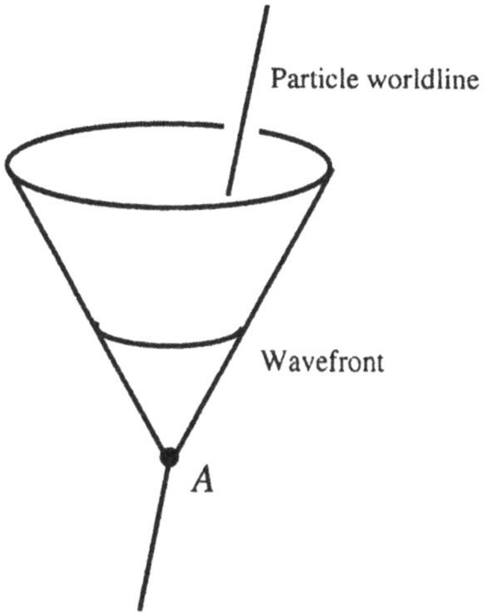

Figure 5.1 The future light-cone of A

up of the events that happen after A that can be reached from A by travelling at the speed of light; and the *past light-cone*, made up of events that happen before A, from which A can be reached by travelling at the speed of light.

The future light-cone of A is illustrated in the space–time diagram Figure 5.1, where, as always, time runs up the page. The sections of the cone on which t is a positive constant are the spherical 'wave-fronts'

$$x^2 + y^2 + z^2 = c^2t^2$$

spreading out from the origin with speed c. By the third assumption, all particle worldlines through A must lie inside the light-cone. We shall always draw space–time diagrams so that the generators of the light-cone are at 45° to the vertical.

In Galilean relativity, an event picks out a *hyperplane of simultaneity*, made up of all the events that are simultaneous with the given event. All observers agree on which set this is since they all agree on which pairs of events are simultaneous. In Einstein's relativity, simultaneity is relative and the $t =$ constant hyperplanes through a given event have no invariant significance. They are different in different inertial coordinate systems. On the other hand, all observers agree on the position in space–time of the light-cone, so the cones associated with each event determine an invariant structure on space–time—a structure that was first made explicit by H. Minkowski. Thus the space–time of special relativity is called *Minkowski space*.

As $c \to 0$, the light-cone degenerates into a $t =$ constant hyperplane through A, and so we recover the Galilean picture in the limit.

5.3 The Lorentz Transformation in Four Dimensions

Consider two events E_1 and E_2 with coordinates t_1, x_1, y_1, z_1 and t_2, x_2, y_2, z_2 in the first coordinate system set up by O; and with coordinates t'_1, x'_1, y'_1, z'_1 and t'_2, x'_2, y'_2, z'_2 in the second coordinate system set up by O'.

The two events lie on the worldline of a photon if and only if

$$c^2(t_2 - t_1)^2 - (x_2 - x_1)^2 - (y_2 - y_1)^2 - (z_2 - z_1)^2 = 0\,, \tag{5.3}$$

since this is the same as the condition $D = cT$, where

$$D = \sqrt{(x_2 - x_1)^2 - (y_2 - y_1)^2 - (z_2 - z_1)^2}$$

is the distance between them and $T = t_2 - t_1$ is time interval between them. Now 'lying on the worldline of a photon' is a property that make sense independently of any choice of coordinate system on space–time. So if $D = cT$ according to one observer, then it must also be true according to the other. Therefore (5.3) holds if and only if

$$c^2(t'_2 - t'_1)^2 - (x'_2 - x'_1)^2 - (y'_2 - y'_1)^2 - (z'_2 - z'_1)^2 = 0\,. \tag{5.4}$$

This statement can be written in a more compact form by putting

$$X = \begin{pmatrix} ct_2 \\ x_2 \\ y_2 \\ z_2 \end{pmatrix} - \begin{pmatrix} ct_1 \\ x_1 \\ y_1 \\ z_1 \end{pmatrix} \qquad \text{and} \qquad X' = \begin{pmatrix} ct'_2 \\ x'_2 \\ y'_2 \\ z'_2 \end{pmatrix} - \begin{pmatrix} ct'_1 \\ x'_1 \\ y'_1 \\ z'_1 \end{pmatrix}$$

and by defining g to be the 4×4 diagonal matrix

$$g = \begin{pmatrix} 1 & 0 & 0 & 0 \\ 0 & -1 & 0 & 0 \\ 0 & 0 & -1 & 0 \\ 0 & 0 & 0 & -1 \end{pmatrix}.$$

Then Equations (5.3) and (5.4) become, respectively,

$$X^{\mathrm{t}} g X = 0 \qquad \text{and} \qquad X'^{\mathrm{t}} g X' = 0\,,$$

where 't' denotes matrix transpose. Since the coordinates of the two events are related by (5.1), we have

$$X = LX'\,.$$

Therefore the equivalence of (5.3) and (5.4) can be stated as follows. For $X \in \mathbb{R}^4$ (with the prime dropped),

$$X^{\mathrm{t}} g X = 0 \qquad \text{if and only if} \qquad X^{\mathrm{t}} L^{\mathrm{t}} g L X = 0\,.$$

Lemma 5.1

Suppose that G is a 4×4 symmetric matrix such that $X^{\mathrm{t}}GX = 0$ whenever $X^{\mathrm{t}}gX = 0$. Then $G = \alpha g$ for some $\alpha \in \mathbb{R}$.

Proof

Write

$$G = \begin{pmatrix} \alpha & a^{\mathrm{t}} \\ a & S \end{pmatrix}$$

where α is a real number, S is a 3×3 symmetric matrix, and a is a column vector of length three. Then by taking $X^{\mathrm{t}} = (1, u, v, w)$, we have

$$\alpha + 2a^{\mathrm{t}} \begin{pmatrix} u \\ v \\ w \end{pmatrix} + (u\,v\,w)S \begin{pmatrix} u \\ v \\ w \end{pmatrix} = 0$$

whenever $u^2 + v^2 + w^2 = 1$. But this last condition on u, v, w is the equation of the unit sphere with centre at the origin; therefore (5.3) must be as well. Hence $a = 0$ and $S = -\alpha I$, where I is the 3×3 identity matrix. The lemma follows. □

Proposition 5.2

$g = L^{\mathrm{t}}gL$.

Proof

It follows from the lemma that $L^{\mathrm{t}}gL = \alpha g$ for some $\alpha \in \mathbb{R}$, which must be nonzero since the coordinate transformation must be nonsingular. Hence

$$L^{-1} = \alpha^{-1} g L^{\mathrm{t}} g$$

since $g^{-1} = g$. Therefore the top left entry in L^{-1} is γ/α, where γ is the top left entry in L. But we deduced from our relativity assumption that the top left entries in L and L^{-1} are equal, so $\alpha = 1$. □

It follows from the proposition that $L^{-1} = gL^{\mathrm{t}}g$ and hence that $LgL^{\mathrm{t}} = g$ since $g^2 = 1$.

5.4 The Standard Lorentz Transformation

Suppose that O is moving along the x'-axis in the coordinates of O' and that O' is moving along the x-axis in the coordinates of O; and suppose further that they both take the origin of their coordinate systems to be the event at which they pass each other. Then $C = 0$ and the t, x and t', x' coordinates are related by (4.1), with $\gamma = \gamma(u)$. Hence

$$L = \begin{pmatrix} \gamma & \gamma u/c & p & q \\ \gamma u/c & \gamma & r & s \\ P & Q & a & b \\ R & S & c & d \end{pmatrix}.$$

From $L^{\mathrm{t}} g L = g$, we obtain

$$\begin{aligned} \gamma^2 - \gamma^2 u^2/c^2 - P^2 - R^2 &= 1 \\ \gamma^2 u^2/c^2 - \gamma^2 - Q^2 - S^2 &= -1\,. \end{aligned}$$

But $\gamma^2 - \gamma^2 u^2/c^2 = 1$. Hence P, Q, R, and S are all zero. Similarly, from $L^{\mathrm{t}} g L = g$, we get that p, q, r, and s are also zero, and then

$$A = \begin{pmatrix} a & b \\ c & d \end{pmatrix}$$

is an orthogonal matrix.

Although the direction of the x-axis in the coordinate system set up by O is fixed by the condition that O' should be travelling along the x-axis, O is still free to make rotations about the x-axis. By making an orthogonal transformation of the y and z coordinates by A^{-1}, it can be can arranged without loss of generality that $A = 1$. We then have

$$\begin{pmatrix} ct \\ x \\ y \\ z \end{pmatrix} = \begin{pmatrix} \gamma & \gamma u/c & 0 & 0 \\ \gamma u/c & \gamma & 0 & 0 \\ 0 & 0 & 1 & 0 \\ 0 & 0 & 0 & 1 \end{pmatrix} \begin{pmatrix} ct' \\ x' \\ y' \\ z' \end{pmatrix}, \tag{5.5}$$

where $\gamma = \gamma(u) = 1/\sqrt{1 - u^2/c^2}$. This is the *standard Lorentz transformation* or *boost* with velocity u. The t, x coordinates transform as before.

Exercise 5.1

Check directly that if L_u is the standard Lorentz transformation matrix with velocity $u < c$, then $L_u^{\mathrm{t}} g L_u = g$.

5.5 The General Lorentz Transformation

In deriving the standard Lorentz transformation, we made assumptions about the relative orientations of the spatial axes of the two coordinate systems. If we drop these, but still assume that O' is moving directly away from O, then we must combine (5.5) with an orthogonal transformation of the x, y, z coordinates and an orthogonal transformation of the x', y', z' coordinates. The result is

$$\begin{pmatrix} ct \\ x \\ y \\ z \end{pmatrix} = L \begin{pmatrix} ct' \\ x' \\ y' \\ z' \end{pmatrix} \tag{5.6}$$

with

$$L = \begin{pmatrix} 1 & 0 \\ 0 & H \end{pmatrix} L_u \begin{pmatrix} 1 & 0 \\ 0 & K^{\mathrm{t}} \end{pmatrix} \tag{5.7}$$

where H and K are 3×3 proper orthogonal matrices, and L_u is the standard Lorentz transformation matrix with velocity u, for some $u < c$. A transformation of the form (5.6) is called a *proper orthochronous Lorentz transformation*. We include the special case $u = 0$, where L_u is the identity and L reduces to a spatial rotation; and the case $H = K = I$, where L is itself a standard Lorentz transformation.

It is conventional to label the entries in L as

$$L = \begin{pmatrix} L^0{}_0 & L^0{}_1 & L^0{}_2 & L^0{}_3 \\ L^1{}_0 & L^1{}_1 & L^1{}_2 & L^1{}_3 \\ L^2{}_0 & L^2{}_1 & L^2{}_2 & L^2{}_3 \\ L^3{}_0 & L^3{}_1 & L^3{}_2 & L^3{}_3 \end{pmatrix} \tag{5.8}$$

using upper and lower indices for reasons that derive from tensor analysis.

Proposition 5.3

If L is a proper orthochronous Lorentz transformation, then

(1) $L^{-1} = gL^{\mathrm{t}}g$.

(2) $L^0{}_0 > 0$.

(3) $\det(L) = 1$.

Proof

The second statement, (2), follows from the fact that $L^0{}_0 = \gamma(u)$. Note that $g^2 = I$ and that (1) and (3) hold both for the standard Lorentz transformation and for any L of the special form

$$L = \begin{pmatrix} 1 & 0 \\ 0 & H \end{pmatrix}, \qquad H \in \mathrm{SO}(3). \tag{5.9}$$

Hence they also hold for any proper orthochronous transformation. □

The positivity of $L^0{}_0$ ensures that t is an increasing function of t'; and the positivity in addition of $\det(L)$ ensures that the 'handedness' of the two sets of spatial axes are the same.

The converse proposition is also true, that is, if L satisfies (1), (2), and (3), then there exist $H, K \in \mathrm{SO}(3)$ such that (5.7) holds, with L_u the standard Lorentz transformation for some u ($|u| < c$). The proof is an exercise. So we can make the following alternative definition.

Definition 5.4

The transformation (5.6) is a *Lorentz transformation* if $L^{-1} = gL^{\mathrm{t}}g$; it is an *orthochronous* Lorentz transformation if in addition $L^0{}_0 > 0$; and it is a *proper* orthochronous Lorentz transformation if it is further true that $\det L = 1$.

Finally, if we drop the condition that O' should be moving directly away from O, then we can combine (5.6) with a spatial translation and a change in the origin of the time coordinate. The result is

$$\begin{pmatrix} ct \\ x \\ y \\ z \end{pmatrix} = L \begin{pmatrix} ct' \\ x' \\ y' \\ z' \end{pmatrix} + C \tag{5.10}$$

where L is a proper orthochronous Lorentz transformation matrix and C is a constant column vector. Equation (5.10) is an *inhomogeneous Lorentz transformation* or, alternatively, a *Poincaré transformation.*

In some contexts, it is useful to put $x^0 = ct$, $x^1 = x$, $x^2 = y$, $x^3 = z$, and so on, and to write (5.10) as

$$x^a = \sum_{b=0}^{3} L^a{}_b x'^b + C^a \qquad (a = 0, 1, 2, 3) \tag{5.11}$$

or, with a summation convention for the repeated index b,

$$x^a = L^a{}_b x'^b + C^a .$$

The positioning of the indices takes a little time to become familiar; it is important in tensor analysis.

5.6 Euclidean Space and Minkowski Space

The four-dimensional space–time of special relativity has a geometry that is analogous in many ways to that of three-dimensional Euclidean space. In Euclidean space, for example, two systems of Cartesian coordinates are related by the transformation

$$\begin{pmatrix} x \\ y \\ z \end{pmatrix} = H \begin{pmatrix} x' \\ y' \\ z' \end{pmatrix} + C \tag{5.12}$$

where $H^{-1} = H^{\mathrm{t}}$ and C is a constant column vector of length three. In Minkowski space, the analogous transformation is the inhomogeneous transformation (5.10) relating two inertial coordinate systems. The orthogonality property of H ($H^{-1} = H^{\mathrm{t}}$) is replaced by the corresponding *pseudo-orthogonality* property of L

$$L^{-1} = (gLg)^{\mathrm{t}}. \tag{5.13}$$

That is, the inverse of L is obtained by changing the signs of the entries $L^0{}_1$, $L^0{}_2$, $L^0{}_3$, $L^1{}_0$, $L^2{}_0$, and $L^3{}_0$, and then by transposing the resulting matrix.

In Euclidean geometry, the squared distance

$$(x_1 - x_2)^2 + (y_1 - y_2)^2 + (z_1 - z_2)^2 \tag{5.14}$$

between (x_1, y_1, z_1) and (x_2, y_2, z_2) is invariant—it is the same in any Cartesian coordinate system; in Minkowski space, the corresponding quantity is the invariant

$$c^2(t_1 - t_2)^2 - (x_1 - x_2)^2 - (y_1 - y_2)^2 - (z_1 - z_2)^2 \tag{5.15}$$

associated with two space–time events (its invariance is established below).

5.7 Four-vectors

The analogies are not complete since there is nothing in Euclidean geometry corresponding to the distinction between orthochronous ($L^0{}_0 > 0$) and non-orthochronous ($L^0{}_0 < 0$) transformations; and they are sometimes misleading since the significance of the minus signs in the pseudo-orthogonality condition can easily be overlooked. But they do suggest that it would be useful to consider space–time vectors, defined by adapting the transformation law for the components of a vector in space.

In three-dimensional Euclidean space, a vector $\boldsymbol{X}$ has three components that transform under (5.12) by

$$\begin{pmatrix} X_1 \\ X_2 \\ X_3 \end{pmatrix} = H \begin{pmatrix} X_1' \\ X_2' \\ X_3' \end{pmatrix},$$

where X_1, X_2, X_3 are the components in the x, y, z coordinate system and X_1', X_2', X_3' are the components in the x', y', z' coordinate system. Although it is not usual to do so, one can use this as the *definition* of a vector; a vector is then characterized as an object that associates a set of components with each Cartesian coordinate system, subject to this transformation rule.

To avoid confusion, vectors in space–time will be called four-vectors and ordinary vectors in Euclidean space will be called three-vectors, although both will be shortened to 'vector' when there is no possibility of ambiguity.

Definition 5.5

A *four-vector* is an object X that associates an element (X^0, X^1, X^2, X^3) of $\mathbb{R}^4$ with each inertial coordinate system. The X^as $(a = 0, 1, 2, 3)$ are called the *components* of X. They are required to have the property that if two inertial coordinate systems are related by (5.10) then the components X^a in the first (unprimed) system are related to components X'^a in the second (primed) by

$$\begin{pmatrix} X^0 \\ X^1 \\ X^2 \\ X^3 \end{pmatrix} = L \begin{pmatrix} X'^0 \\ X'^1 \\ X'^2 \\ X'^3 \end{pmatrix}. \tag{5.16}$$

This rather awkward definition says no more than that a four-vector is an object with four components X^0, X^1, X^2, X^3 and that the components transform under an inhomogeneous Lorentz transformation of the coordinates by the associated linear transformation—the same transformation as the coordinates, but without the constant column vector. The important point to remember is which way round the transformation goes (look carefully at the relationship between (5.10) and (5.16)).

As with three-vectors, one can add four-vectors and take scalar multiples. Thus

- The four-vector $X+Y$ has components $(X^0+Y^0, X^1+Y^1, X^2+Y^2, X^3+Y^3)$.
- The four-vector λX $(\lambda \in \mathbb{R})$ has components $(\lambda X^0, \lambda X^1, \lambda X^2, \lambda X^3)$.

The set of all four-vectors is a four-dimensional vector space.

A key example is the *displacement vector* X from an event E_1 to an event E_2. If the events have respective coordinates t_1, x_1, y_1, z_1 and t_2, x_2, y_2, z_2 in some inertial coordinate system, then the displacement vector X from E_1 to E_2 has components

$$X^0 = ct_2 - ct_1, \quad X^1 = x_2 - x_1, \quad X^2 = y_2 - y_1, \quad X^3 = z_2 - z_1 .$$

Under the inhomogeneous transformation

$$\begin{pmatrix} ct \\ x \\ y \\ z \end{pmatrix} = L \begin{pmatrix} ct' \\ x' \\ y' \\ z' \end{pmatrix} + C, \tag{5.17}$$

the X^as transform according to (5.16) where now $X'^0 = ct'_2 - ct'_1$ and so on. Thus the components have the four-vector transformation property.

If E_3 is a third event, and Y is the displacement vector from E_2 to E_3 and Z is the displacement vector from E_1 to E_3, then we have the *triangle relation*

$$Z = X + Y$$

(Figure 5.2). The displacement vector between two events is analogous to the vector $\overrightarrow{AB}$ between two points A, B in Euclidean space.

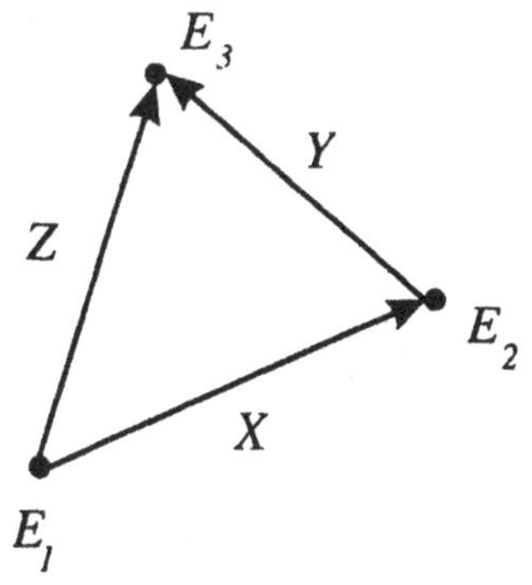

Figure 5.2 The triangle relation

5.8 Temporal and Spatial Parts

If two inertial observers O and O' are at rest relative to each other, then their time axes in space–time will be aligned and their inertial coordinate systems will be related by

$$\begin{pmatrix} ct \\ x \\ y \\ z \end{pmatrix} = \begin{pmatrix} 1 & 0 \\ 0 & H \end{pmatrix} \begin{pmatrix} ct' \\ x' \\ y' \\ z' \end{pmatrix} + C \,, \tag{5.18}$$

where H is a 3×3 proper orthogonal matrix and C is a column vector—that is, by a rotation of the x, y, z axes, combined with a translation in space and time. In this special case, the components of a four-vector X in the two systems are related by

$$X^0 = X'^0, \qquad \begin{pmatrix} X^1 \\ X^2 \\ X^3 \end{pmatrix} = H \begin{pmatrix} X'^1 \\ X'^2 \\ X'^3 \end{pmatrix} . \tag{5.19}$$

The time component is the same, while the three spatial components X^1, X^2, X^3 behave as the components of a three-vector $\boldsymbol{x}$. So we can decompose X into a *temporal part* X^0 and a *spatial part* $\boldsymbol{x}$ in a way that depends only on the velocity of the observer and not on the particular choice of origin and orientation of the spatial coordinate axes. The decomposition will be unchanged by the transformation between the coordinate systems of two observers at rest relative to each other. By contrast, under a general transformation between the inertial coordinate systems of two observers in relative motion, the direction of the time axis changes and the temporal and spatial parts are mixed up.

We shall write

$$X = (X^0, X^1, X^2, X^3)$$

as shorthand for 'X has components X^0, X^1, X^2, X^3' in a particular inertial coordinate system and

$$X = (\xi, \boldsymbol{x})$$

for 'X has temporal part ξ and spatial part $\boldsymbol{x}$' relative to a particular choice for direction in space–time of the t-axis.

5.9 The Inner Product

In Euclidean geometry, the distance between two points is determined by the dot product, which is an *inner product* on the space of three-vectors. If A, B are points in space and if $\boldsymbol{x}$ is the vector from A to B, then the distance from

A to B is $\sqrt{\boldsymbol{x}\,.\,\boldsymbol{x}}$. In a Cartesian coordinate system, the dot product of two vectors $\boldsymbol{a}$ and $\boldsymbol{b}$ is given in terms of their components by

$$\boldsymbol{a}\,.\,\boldsymbol{b} = a_1 b_1 + a_2 b_2 + a_3 b_3\,.$$

The right-hand side is independent of the choice of coordinates since the components of three-vectors in different Cartesian coordinate systems are related by an orthogonal transformation. The inner product is *positive definite* in the sense that

$$\boldsymbol{a}\,.\,\boldsymbol{a} \geq 0\,,$$

with equality only if $\boldsymbol{a} = 0$.

The pseudo-orthogonality of Lorentz transformations leads to an analogous *indefinite* inner product on the space of four-vectors. That is, it has all the properties of an inner product, except that it is not positive definite.

Definition 5.6

The *inner product* $g(X, Y)$ of two four vectors X and Y is the real number

$$g(X,Y) = X^0Y^0 - X^1Y^1 - X^2Y^2 - X^3Y^3\,,$$

where X^a, Y^a, $a = 0,1,2,3$, are the components of X and Y in an inertial coordinate system.

Proposition 5.7

The inner product of two four-vectors X and Y is independent of the choice of inertial coordinate system.

Proof

We can write the definition of $g(X,Y)$ in matrix form

$$g(X,Y) = \begin{pmatrix} X^0 & X^1 & X^2 & X^3 \end{pmatrix} g \begin{pmatrix} Y^0 \\ Y^1 \\ Y^2 \\ Y^3 \end{pmatrix}, \tag{5.20}$$

where g is the diagonal matrix introduced in §5.3. Now use Equation (5.16) to express X and Y in terms of their components X'^a, Y'^a in a second inertial coordinate system. Then the right-hand side of (5.20) becomes

$$\begin{pmatrix} X'^0 & X'^1 & X'^2 & X'^3 \end{pmatrix} L^{\mathrm{t}} g L \begin{pmatrix} Y'^0 \\ Y'^1 \\ Y'^2 \\ Y'^3 \end{pmatrix} = \begin{pmatrix} X'^0 & X'^1 & X'^2 & X'^3 \end{pmatrix} g \begin{pmatrix} Y'^0 \\ Y'^1 \\ Y'^2 \\ Y'^3 \end{pmatrix},$$

since $L^t g L = g$. So the result is the same, whether we calculate $g(X,Y)$ in the new coordinate system or the old. □

The inner product is a symmetric bilinear form of the space of four-vectors. That is,

- it is symmetric since $g(X,Y) = g(Y,X)$ for all four-vectors X and Y;
- it is linear in each argument since

$$g(\lambda X + \mu Y, Z) = \lambda g(X,Z) + \mu g(Y,Z)$$

for all four-vectors X, Y, Z and all scalars λ and μ.

Note, however, that

$$g(X,X) = \xi^2 - \boldsymbol{x}.\boldsymbol{x},$$

where ξ and $\boldsymbol{x}$ are the temporal and spatial parts of X. So $g(X,X)$ can be positive or negative, as $\xi > |\boldsymbol{x}|$ or $\xi < |\boldsymbol{x}|$; and it can be zero even if $X \neq 0$. Strictly speaking, therefore, we should not describe g as an 'inner product', since the standard terminology in linear algebra requires an inner product to be positive definite as well as symmetric and bilinear; but the term 'indefinite inner product' is hardly more satisfactory, and is certainly more clumsy. We shall follow a common path in departing from algebraic convention and in dropping the qualification 'indefinite'.

Aside

The transformation rule for four-vector components can be written more compactly as

$$X^a = L^a{}_b X'^b \quad \text{when} \quad x^a = L^a{}_b x'^b + C^a, \tag{5.21}$$

where in both equations the repetition of the index b implies a sum over $b = 0, 1, 2, 3$; see Equation (5.11). Similarly, the invariant $g(X,Y)$ can be written

$$g(X,Y) = g_{ab} X^a Y^b \tag{5.22}$$

where the g_{ab}s are the entries in the matrix g. That is,

$$g_{00} = 1, \qquad g_{11} = g_{22} = g_{33} = -1,$$

and $g_{ab} = 0$ when $a \neq b$. In (5.22), there are two summations over the repeated indices $a, b = 0, 1, 2, 3$.

A further notational device in common use is to put $X_a = g_{ab} X^b$, again with a summation over b. Then $X_0 = X^0$, $X_1 = -X^1$, $X_2 = -X^2$, $X_3 = -X^3$ and

$$g(X,Y) = X_a Y^a = X_0 Y^0 + X_1 Y^1 + X_2 Y^2 + X_3 Y^3. \tag{5.23}$$

The operation of forming the X_as from the components X^a of X is called 'lowering the index'. The conventions for the positioning of indices are such that summations are always over one lower index and one upper index.

5.10 Classification of Four-vectors

The fact that the inner product on four-vectors is not positive definite means that it is possible to distinguish between different types of four-vector according to the sign of the invariant $g(X, X)$.

Definition 5.8

A four-vector X is said to be *timelike*, *spacelike*, or *null* as $g(X, X) > 0$, $g(X, X) < 0$, or $g(X, X) = 0$. Two four-vectors X and Y are *orthogonal* if $g(X, Y) = 0$.

Example 5.9

The four-vectors with components

$$(1, 0, 0, 0), \quad (0, 1, 0, 0), \quad \text{and} \quad (1, 1, 0, 0)$$

in some inertial coordinate system are respectively timelike, spacelike, and null. Note that a null four-vector need not be the zero four-vector.

A four-vector whose spatial part vanishes in some inertial coordinate system must be timelike; and a four-vector whose temporal part vanishes in some inertial coordinate system must be spacelike. The converses of these statements are given by the following proposition.

Proposition 5.10

If X is timelike, then there exists an inertial coordinate system in which $X^1 = X^2 = X^3 = 0$. If X is spacelike, then there exists and inertial coordinate system in which $X^0 = 0$.

Proof

Consider the components X^0, X^1, X^2, X^3 of X in an inertial coordinate system

t, x, y, z. By rotating the spatial axes to make the x-axis parallel to the spatial part of X, we can ensure that $X^2 = X^3 = 0$.

We can then make the third spatial component vanish by making a standard Lorentz transformation chosen so that

$$\gamma(u)\begin{pmatrix} 1 & u/c \\ u/c & 1 \end{pmatrix}\begin{pmatrix} X^0 \\ X^1 \end{pmatrix} = \begin{pmatrix} * \\ 0 \end{pmatrix}. \tag{5.24}$$

That is, we choose u such that $|u| < c$ and

$$uX^0/c + X^1 = 0. \tag{5.25}$$

This is possible because X is timelike, and so $|X^1/X^0| < 1$. Similarly, if X is spacelike, then $|X^0/X^1| < 1$ and we can find u with $|u| < c$ to make $X^0 = 0$. □

In the case of timelike and null vectors (but *not* spacelike vectors), the sign of the time component X^0 is invariant.

Proposition 5.11

Suppose that $X \neq 0$ is timelike or null. If $X^0 > 0$ in some inertial coordinate system, then $X^0 > 0$ in every inertial coordinate system.

Proof

Since rotations do not alter X^0, it is sufficient to consider what happens to X^0 under a standard Lorentz transformation. So suppose that

$$\begin{pmatrix} X^0 \\ X^1 \\ X^2 \\ X^3 \end{pmatrix} = \begin{pmatrix} \gamma & \gamma u/c & 0 & 0 \\ \gamma u/c & \gamma & 0 & 0 \\ 0 & 0 & 1 & 0 \\ 0 & 0 & 0 & 1 \end{pmatrix}\begin{pmatrix} X'^0 \\ X'^1 \\ X'^2 \\ X'^3 \end{pmatrix}, \tag{5.26}$$

where $\gamma = \gamma(u)$. Then

$$X^0 = \gamma X'^0\left(1 + \frac{u}{c}\frac{X'^1}{X'^0}\right). \tag{5.27}$$

Since $|u/c| < 1$ and $|X'^1/X'^0| < 1$, the expression in brackets is positive. It follows that X^0 and X'^0 have the same sign (both are nonzero). □

Definition 5.12

A timelike or null vector X is said to be *future-pointing* if $X^0 > 0$ in some (and hence every) inertial coordinate system, and *past-pointing* if $X^0 < 0$.

We cannot make an analogous definition for spacelike vectors since the sign of the time component of a spacelike vector is different in different inertial coordinate systems. The space of four-vectors is illustrated in Figure 5.3, where

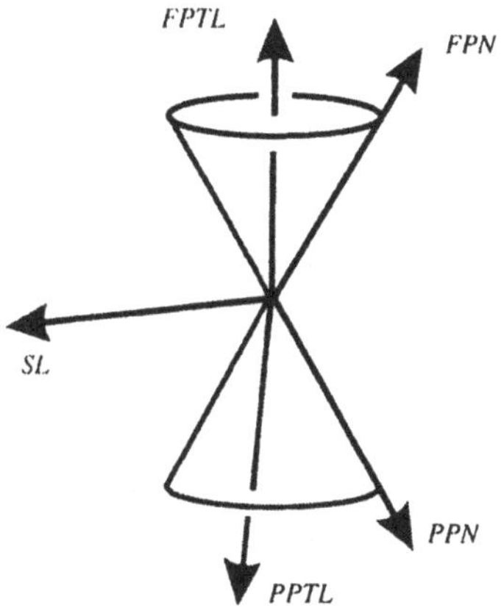

Figure 5.3 The space of 4-vectors

the time axis is vertical and one spatial dimension is suppressed, and where 'FPTL' denotes 'future-pointing timelike', and so on. The null vectors lie on the cone

$$(X^0)^2 - (X^1)^2 - (X^2)^2 - (X^3)^2 = 0\,. \tag{5.28}$$

5.11 Causal Structure of Minkowski Space

In the case of displacement four-vectors, the classification has a direct interpretation in terms of the *causal structure* of Minkowski space. Suppose that E and F are events and that X is the displacement four-vector from E to F. In studying the *causal relationship* between E and F, we are interested in whether it is possible for some physical process that happens at E to influence what happens at F, or the other way around. If it is possible, then we are also interested in whether effects can be transmitted from one event to the other at less than the speed of light or only at exactly the speed of light. We exclude, of course, the possibility of transmission faster than the speed of light.

The temporal part of X in an inertial coordinate system is the time from E to F multiplied by c; the spatial part is the three-vector from the point where

E happens to the point where F happens. There various possibilities.

- The displacement X is spacelike. In this case, it is impossible to get from E to F without travelling faster than light, so F lies outside the light cone of E, and vice versa. By Proposition 5.10, there exists an inertial coordinate system in which $X^0 = 0$; that is, in which E and F are simultaneous. If s denotes the distance from E to F in such coordinates, then

$$g(X, X) = -s^2 .$$

There also exist inertial coordinate systems in which E happens before F, and inertial coordinate systems in which E happens after F. It is for this reason that the prohibition on faster-than-light transmission is required for the consistency of the theory with commonsense ideas about causality.

- The displacement X is timelike. In this case there exists an inertial coordinate system in which $X^1 = X^2 = X^3 = 0$; that is in which E and F happen at the same place. If τ denotes the time from E to F in such coordinates, then

$$g(X, X) = c^2\tau^2 .$$

If X is future-pointing, then $\tau > 0$ and F happens after E in every inertial coordinate system. If X is past-pointing, then $\tau < 0$ and F happens before E in every inertial coordinate system.

- The displacement X is null. Then E and F lie on the worldline of a photon. If X is future-pointing (past-pointing), then F happens after (before) E in every inertial coordinate system.

5.12 Invariant Operators

In Euclidean space, the three partial derivatives

$$\partial_x = \frac{\partial}{\partial x}, \quad \partial_y = \frac{\partial}{\partial y}, \quad \partial_z = \frac{\partial}{\partial z}$$

transform as the components of a vector operator $\boldsymbol{\nabla}$. In other words, if we make the coordinate transformation

$$\begin{pmatrix} x \\ y \\ z \end{pmatrix} = H \begin{pmatrix} x' \\ y' \\ z' \end{pmatrix} + C \tag{5.29}$$

where H is a proper orthogonal matrix and C is a constant column vector, then

$$\begin{pmatrix} \partial_x \\ \partial_y \\ \partial_z \end{pmatrix} = H \begin{pmatrix} \partial_{x'} \\ \partial_{y'} \\ \partial_{z'} \end{pmatrix} . \tag{5.30}$$

By making $\boldsymbol{\nabla}$ act on a scalar field f or a vector field $\boldsymbol{X}$, we can form the familiar differential operators

$$\begin{aligned} \operatorname{grad} f &= \boldsymbol{\nabla} f \\ \operatorname{div} \boldsymbol{X} &= \boldsymbol{\nabla} . \boldsymbol{X} \\ \operatorname{curl} \boldsymbol{X} &= \boldsymbol{\nabla} \wedge \boldsymbol{X} . \end{aligned}$$

They are invariant in the sense that they are the same regardless of the Cartesian coordinate system in which they are applied, provided that the axes are right-handed.

It is not immediately obvious that (5.30) follows from (5.29), since the proof makes use of the orthogonality of H—one cannot make this step when orthogonality is dropped and H is allowed to be a general matrix. However, the pseudo-orthogonality of Lorentz transformations enables us to construct four-vector versions of grad and div in much the same way, although the four-dimensional version of curl is a more complicated object since the existence of the vector product in Euclidean space is a special feature of three-dimensional space. We shall introduce the *four-gradient*, which sends a function on space–time to a four-vector field—a four-vector that varies from event to event—and the *four-divergence*, which sends a four-vector field to a scalar function. To define them, we need the following.

Let t, x, y, z be an inertial coordinate system and put

$$\partial_a = \partial/\partial x^a, \qquad a = 0, 1, 2, 3,$$

where $x^0 = ct$, $x^1 = x$, $x^2 = y$, $x^3 = z$. For a function f of the coordinates, put

$$\partial f = (\partial_0 f, \partial_1 f, \partial_2 f, \partial_3 f) .$$

Now consider an inhomogeneous Lorentz transformation

$$\begin{pmatrix} x^0 \\ x^1 \\ x^2 \\ x^3 \end{pmatrix} = L \begin{pmatrix} x'^0 \\ x'^1 \\ x'^2 \\ x'^3 \end{pmatrix} + C , \tag{5.31}$$

and define $\partial' f$ in the same way by putting $\partial'_0 f = \partial f/\partial x'^0$, and so on. Then we have the following.

Lemma 5.13

For any function f on space–time, $\partial' f = \partial f L$.

Proof

If we label the entries in L as in (5.8), then $\partial x^a/\partial x'^b = L^a{}_b$. By the chain rule

$$\frac{\partial f}{\partial x'^b} = \sum_{a=0}^{3} \frac{\partial f}{\partial x^a}\frac{\partial x^a}{\partial x'^b} = \sum_{a=0}^{3} \frac{\partial f}{\partial x^a} L^a{}_b\,,$$

which, in matrix notation, is $\partial' f = \partial f L$. □

This is very close to the transformation rule for a vector (it is in fact the transformation rule for a *covector*—see §5.14). By using the pseudo-orthogonality relation $L^{-1} = gL^{\mathrm{t}}g$, we can rewrite it in the form

$$\partial f g = \partial' f g L^{\mathrm{t}} \qquad \text{or} \qquad (\partial f g)^{\mathrm{t}} = L(\partial' f g)^{\mathrm{t}}\,,$$

which is the four-vector transformation rule. So we have shown that

$$(\partial f g)^{\mathrm{t}} = \begin{pmatrix} \partial_0 f \\ -\partial_1 f \\ -\partial_2 f \\ -\partial_3 f \end{pmatrix} \tag{5.32}$$

transforms as a four-vector.

Definition 5.14

The *four-gradient* Grad f of a function f on space–time is the four-vector with components (5.32).

Another way to express the transformation rule of the four-gradient is to say that

$$\frac{1}{c}\frac{\partial}{\partial t}, \quad -\frac{\partial}{\partial x}, \quad -\frac{\partial}{\partial y}, \quad \text{and} \quad -\frac{\partial}{\partial z}\,,$$

transform as the components of a four-vector operator, just as do the components of ∇ in Euclidean space. Given a four-vector field X, that is a four-vector that varies from event to event, we can form an invariant scalar Div X by taking the inner product of the four-vector operator with X. The result is the following.

Definition 5.15

The *four-divergence* of a four-vector field X is the function

$$\mathrm{Div}\, X = \frac{1}{c}\frac{\partial X^0}{\partial t} + \frac{\partial X^1}{\partial x} + \frac{\partial X^2}{\partial y} + \frac{\partial X^3}{\partial z}.$$

Proposition 5.16

The four-divergence is invariant under change of inertial coordinate system.

Proof

Under the transformation (5.31), we have

$$\begin{pmatrix}\partial_0 & \partial_1 & \partial_2 & \partial_3\end{pmatrix} = \begin{pmatrix}\partial_0' & \partial_1' & \partial_2' & \partial_3'\end{pmatrix} L^{-1}$$

since the components of L^{-1} are constants; and also

$$\begin{pmatrix}X^0 \\ X^1 \\ X^2 \\ X^3\end{pmatrix} = L \begin{pmatrix}X'^0 \\ X'^1 \\ X'^2 \\ X'^3\end{pmatrix}.$$

It follows that

$$\mathrm{Div}\, X = \begin{pmatrix}\partial_0 & \partial_1 & \partial_2 & \partial_3\end{pmatrix}\begin{pmatrix}X^0 \\ X^1 \\ X^2 \\ X^3\end{pmatrix} = \begin{pmatrix}\partial_0' & \partial_1' & \partial_2' & \partial_3'\end{pmatrix}\begin{pmatrix}X'^0 \\ X'^1 \\ X'^2 \\ X'^3\end{pmatrix},$$

and hence that the four-divergence is invariant. □

Exercise 5.2

Show that if X is a four-vector field and f is a function on space–time, then

$$\mathrm{Div}\,(fX) = f\,\mathrm{Div}\,(X) + g(\mathrm{Grad}\, f, X).$$

Given a function u on space–time, we can form the four-gradient $\mathrm{Grad}\, u$ and then take the four-divergence. The result is the d'Alembertian, or wave operator, since

$$\Box u = \frac{1}{c^2}\frac{\partial^2 u}{\partial t^2} - \frac{\partial^2 u}{\partial x^2} - \frac{\partial^2 u}{\partial y^2} - \frac{\partial^2 u}{\partial z^2} = \mathrm{Div}\,(\mathrm{Grad}\, u).$$

It follows that the d'Alembertian is an invariant operator; the result of applying it to a function is independent of the choice of inertial coordinates.

Finally, there is also a four-dimensional counterpart of

$$(\boldsymbol{x} \,.\, \boldsymbol{\nabla})\boldsymbol{y}\,,$$

where $\boldsymbol{x}$ and $\boldsymbol{y}$ are three-vector fields. This is the *covariant derivative* $\nabla_X Y$ of a four-vector field Y along a four-vector field X. It is the four-vector field with components

$$X^0\partial_0\begin{pmatrix}Y^0\\Y^1\\Y^2\\Y^3\end{pmatrix}+X^1\partial_1\begin{pmatrix}Y^0\\Y^1\\Y^2\\Y^3\end{pmatrix}+X^2\partial_2\begin{pmatrix}Y^0\\Y^1\\Y^2\\Y^3\end{pmatrix}+X^3\partial_3\begin{pmatrix}Y^0\\Y^1\\Y^2\\Y^3\end{pmatrix}.$$

It is left as an exercise to show that ∇_X is also an invariant operator.

Exercise 5.3

Show that $\nabla_X(fY) = f\nabla_X Y + g(X, \operatorname{Grad} f)Y$, where X, Y are four-vector fields and f is a function on space–time.

5.13 The Frequency Four-vector

We defined real harmonic waves in §3.3. We can most easily understand how such solutions of the wave equation behave under change of coordinates by noting that a real harmonic wave is the real part of a complex solution of the form

$$\psi = A\exp(-\mathrm{i}\Omega)\,, \tag{5.33}$$

where $A = \alpha + \mathrm{i}\beta$ is constant and Ω is a real linear function of the inertial coordinates. Such solutions are characterized by the condition that

$$K = \frac{\mathrm{i}c\operatorname{Grad}\psi}{\psi}$$

should be a real constant four-vector.

Proposition 5.17

Suppose that ψ is a non-zero complex-valued function on space–time such that $K = \mathrm{i}c\operatorname{Grad}\psi/\psi$ is a real constant four-vector. Then $\Box\psi = 0$ if and only if K is null.

Proof

If $\operatorname{Grad}\psi = -\mathrm{i}\psi K/c$ where K is constant, then $\Box\psi = -g(K,K)\psi/c^2$, by Exercise 5.2. The proposition follows. □

Definition 5.18

A *complex harmonic wave* ψ is a complex-valued function on space–time such that $K = \mathrm{i}c\operatorname{Grad}\psi/\psi$ is a real constant null four-vector.

The real part of a complex harmonic wave is a real harmonic wave. Conversely if u is a real harmonic wave, given by (3.11), then $u = \operatorname{Re}\psi$, where ψ is the complex harmonic wave (5.33). In this case, K has temporal and spatial parts

$$K = \omega(1, \boldsymbol{e}),$$

where ω is the frequency and $\boldsymbol{e}$ is the unit vector in the direction of propagation. Note that

$$g(K,K) = \omega^2(1 - \boldsymbol{e} \cdot \boldsymbol{e}) = 0.$$

There is one subtlety here that is worth noting. If $u = \operatorname{Re}\psi$, then also $u = \operatorname{Re}\overline{\psi}$. Thus a real harmonic wave can be written in two distinct ways as the real part of a complex harmonic wave. In one case K is future-pointing, and in the other it is past-pointing. This is an important point in quantum field theory, where it is related to the distinction between particles and anti-particles. But in our classical context we shall avoid it by making the convention that ψ should always be chosen so that K is future-pointing. So we adopt the following.

Definition 5.19

The *frequency four-vector* of a real harmonic wave u is the constant null four-vector $K = \mathrm{i}c\operatorname{Grad}\psi/\psi$, where ψ is the complex harmonic wave chosen so that $u = \operatorname{Re}\psi$ and so that K is future-pointing.

5.14 *Affine Spaces and Covectors

In this chapter we have adopted a traditional definition of four-vectors in terms of the transformation rule for their components. Such a pragmatic approach is useful in explicit calculation, but is not very helpful in making clear their exact nature as mathematical objects.

We can do this by describing the structure of Minkowski space in more precise language. Let $\mathbb{V}$ be a four-dimensional real vector space, on which there is given an indefinite inner product g with signature $+,-,-,-$. This means that there are bases in $\mathbb{V}$—called *pseudo-orthonormal* bases—in which

$$g(X,X) = (X^0)^2 - (X^1)^2 - (X^1)^2 - (X^1)^2, \tag{5.34}$$

where $X \in V$ has components X^0, X^1, X^2, X^3 in the basis.

Minkowski space $\mathbb{M}$ is the *affine space* modelled on $\mathbb{V}$. The full definition is given in §9.1, but the key property is that elements of $\mathbb{V}$ should determine displacements between pairs of events in $\mathbb{M}$. In a similar spirit, Euclidean space is the affine space modelled on the space of three-vectors, with the dot product determining distance.

A four-vector in Minkowski space is simply an element of $\mathbb{V}$. A *covector* α is an element of the dual space $\mathbb{V}^*$. In the dual of a pseudo-orthonormal basis, α has components $\alpha_0, \alpha_1, \alpha_2, \alpha_3$. When the basis is replaced by another pseudo-orthonormal basis, the components of four-vectors and covectors transform by

$$\begin{pmatrix} X^0 \\ X^1 \\ X^2 \\ X^3 \end{pmatrix} = L \begin{pmatrix} X'^0 \\ X'^1 \\ X'^2 \\ X'^3 \end{pmatrix} \quad \text{and} \quad \begin{pmatrix} \alpha_0, \ \alpha_1, \ \alpha_2, \ \alpha_3 \end{pmatrix} = \begin{pmatrix} \alpha'_0, \ \alpha'_1, \ \alpha'_2, \ \alpha'_3 \end{pmatrix} L^{-1}$$

where L is pseudo-orthogonal. In particular, the transformation rule in Lemma 5.13 implies that $\partial_a f$, $a = 0, 1, 2, 3$, are the components of a covector.

EXERCISES

5.4. Let O and O' be two non-accelerating observers whose inertial coordinate systems are related by a proper orthochronous Lorentz transformation

$$\begin{pmatrix} ct \\ x \\ y \\ z \end{pmatrix} = L \begin{pmatrix} ct' \\ x' \\ y' \\ z' \end{pmatrix}.$$

Show that

$$L = \begin{pmatrix} \gamma & -\gamma u'_1/c & -\gamma u'_2/c & -\gamma u'_3/c \\ \gamma u_1/c & * & * & * \\ \gamma u_2/c & * & * & * \\ \gamma u_3/c & * & * & * \end{pmatrix},$$

where $\boldsymbol{u} = (u_1, u_2, u_3)$ is the velocity of O' in the inertial coordinate system of O, $\boldsymbol{u}' = (u'_1, u'_2, u'_3)$ is the velocity of O in the inertial coordinate system of O', and $\gamma = \gamma(u) = \gamma(u')$.

5.5. Which of the following are Lorentz transformation matrices? Which are proper and orthochronous?

$$\begin{pmatrix} \sqrt{2} & 1 & 0 & 0 \\ 1 & \sqrt{2} & 0 & 0 \\ 0 & 0 & 1 & 0 \\ 0 & 0 & 0 & 1 \end{pmatrix}, \qquad \frac{1}{\sqrt{2}}\begin{pmatrix} 2 & 0 & 1 & -1 \\ 1 & 1 & 1 & -1 \\ -1 & 1 & -1 & 1 \\ 0 & 0 & 1 & 1 \end{pmatrix},$$

$$\frac{1}{\sqrt{2}}\begin{pmatrix} -2 & 1 & 0 & -1 \\ -1 & 1 & 1 & -1 \\ 1 & -1 & 1 & 1 \\ 0 & 1 & 0 & 1 \end{pmatrix}, \qquad \frac{1}{\sqrt{2}}\begin{pmatrix} 2 & 1 & 0 & -1 \\ 1 & 1 & 1 & -1 \\ -1 & -1 & 1 & 1 \\ 0 & 1 & 0 & 1 \end{pmatrix}.$$

5.6. Let V be a four-vector. Show that (i) if V is future-pointing timelike, then there exists an inertial coordinate system in which it has components $(a, 0, 0, 0)$, where $a = \sqrt{g(V, V)}$, and (ii) if V is future-pointing null, then there exists an inertial coordinate system in which V has components $(1, 1, 0, 0)$.

5.7. Show that

(i) the sum of two future-pointing timelike four-vectors is future-pointing timelike;

(ii) the sum of two future-pointing null four-vectors is either timelike or null, and is again future-pointing; under what condition is the sum null?

(iii) every four-vector orthogonal to a timelike vector is spacelike.

5.8. Let X and Y be future-pointing timelike four-vectors, and let $Z = X + Y$. Show that

$$\sqrt{g(Z, Z)} \geq \sqrt{g(X, X)} + \sqrt{g(Y, Y)}\,.$$

When does equality hold? What is the analogous statement in Euclidean geometry?

5.9. Let X denote the displacement vector from a fixed origin in space–time, let $\sigma^2 = g(X, X)$ and let f be a function of one variable. Show that

$$\operatorname{Grad} f(\sigma^2) = 2f'(\sigma^2)X, \qquad \operatorname{Div} X = 4\,.$$

Hence show that

$$\frac{1}{g(X, X)} = \frac{1}{c^2t^2 - x^2 - y^2 - z^2}$$

is a solution of the wave equation.

5.10. Show that for any constant complex four-vector A (i.e. for any constant four-vector with complex components), $u = g(X - A, X - A)^{-1}$ is a complex solution of the wave equation, and that its real part is a real solution. By taking the imaginary part of A to be timelike, construct a non-singular, non-constant bounded real solution of the wave equation.

5.11. +Plane harmonic waves of frequencies $1/p$, $1/q$, $1/r$, and $1/s$ are travelling, respectively, in the directions of the (non-unit) vectors $(1, 1, 1)$, $(1, -1, -1)$, $(-1, 1, -1)$, and $(-1, -1, 1)$. Show that there exists an inertial coordinate system in which they have the same frequency if and only if

$$3(p+q-r-s)^2+3(p-q+r-s)^2+3(p-q-r+s)^2 < (p+q+r+s)^2 .$$

6
Relative Motion

6.1 Transformations Between Frames

In Galilean relativity, the motion of a body relative to one inertial frame determines its motion relative to any other inertial frame in a very straightforward way. The motion relative to the first frame combines with that of the first frame relative to the second by vector addition of velocities. In Chapter 4, however, we saw that the rule for combining velocities along a line is different in Einstein's theory, not least because a photon moving at the velocity of light relative to one inertial frame also moves at the velocity of light relative to another.

We now consider the general motion in space of particles, photons, and continuous bodies, and look at the problem of understanding the relationship between the descriptions of the motion in different inertial frames. There are two aspects to this: first, the transformation of the equations for particle worldlines, which gives the relationship between different coordinate descriptions of the motion; second, the relationship between what observers actually *see*, which is more subtle because it involves the trajectories of photons that travel from the body to reach an observer at a particular event. It also raises questions about the relationship between solutions of the wave equation in different coordinate systems.

6.2 Proper Time

First we consider the worldline of a non-accelerating particle. This is a straight line Γ in space–time which lies inside the light cone of any event on Γ. We shall label the events along Γ by using the time τ shown on a clock carried by the particle; τ is also the time coordinate in an inertial coordinate system set up by an observer moving with the particle, relative to whom the particle is at rest.

Definition 6.1

The *proper time* along the worldline of a particle in uniform motion is the time measured in an inertial coordinate system in which the particle is at rest.

Proper time is a natural parameter, analogous to the distance along a line in Euclidean space. It is well-defined up to the addition of a constant, which is determined by the choice of the event $\tau = 0$.

If the particle is at rest in the inertial coordinate system $\tilde{t}, \tilde{x}, \tilde{y}, \tilde{z}$ then its worldline is given by $\tilde{t} = \tau$, with $\tilde{x}$, $\tilde{y}$, $\tilde{z}$ constant. If t, x, y, z is a second inertial coordinate system related to $\tilde{t}, \tilde{x}, \tilde{y}, \tilde{z}$ by the standard Lorentz transformation with velocity v, then

$$t = \gamma(v)\tau + \text{constant} \tag{6.1}$$

along Γ. Since the coordinate time is unchanged by rotation or translation, it follows that in a general inertial coordinate system t, x, y, z,

$$\frac{\mathrm{d}t}{\mathrm{d}\tau} = \gamma(v) = \frac{1}{\sqrt{1 - v^2/c^2}}, \tag{6.2}$$

where v is the speed of the particle. We see again the *time dilation* effect; when $v \neq 0$, a clock carried by the particle, which shows proper time, runs slow relative to the coordinate time t. If two events on Γ are separated by one hour of proper time, then they are separated by $t = \gamma(v) > 1$ hours of coordinate time.

It is important not to be dazzled by the terminology; Equation (6.2) simply gives the relationship between two different ways of measuring time between events, on the one hand by carrying a clock at constant speed between them (τ) and on the other by using Milne's radar method from another location (t). That they give different answers reflects the fact that we have abandoned the uncritical acceptance of the classical view that 'time' has an absolute meaning; in Einstein's theory 'time' is defined in terms of the operations involved in measuring it. There is nothing mysterious about the fact that different operations give different answers.

6.3 Four-velocity

Along the particle worldline, the inertial coordinates are functions of the τ. Put

$$V^0 = c\frac{\mathrm{d}t}{\mathrm{d}\tau}, \quad V^1 = \frac{\mathrm{d}x}{\mathrm{d}\tau}, \quad V^2 = \frac{\mathrm{d}y}{\mathrm{d}\tau}, \quad V^3 = \frac{\mathrm{d}z}{\mathrm{d}\tau}. \tag{6.3}$$

Then we have the following.

Proposition 6.2

The V^as are the components of a four-vector.

Proof

We have to show that the components transform correctly under change of coordinates. Suppose that t', x', y', z' is a second inertial coordinate system which is related to t, x, y, z by an inhomogeneous Lorentz transformation

$$\begin{pmatrix} ct \\ x \\ y \\ z \end{pmatrix} = L \begin{pmatrix} ct' \\ x' \\ y' \\ z' \end{pmatrix} + C. \tag{6.4}$$

Then by differentiating both sides with respect to τ along Γ

$$\begin{pmatrix} V^0 \\ V^1 \\ V^2 \\ V^3 \end{pmatrix} = L \begin{pmatrix} V'^0 \\ V'^1 \\ V'^2 \\ V'^3 \end{pmatrix}, \tag{6.5}$$

where $V'^0 = c\,\mathrm{d}t'/\mathrm{d}\tau$, and so on. This is the four-vector transformation rule. □

Definition 6.3

The four-vector V with components (V^0, V^1, V^2, V^3) is called the *four-velocity* of the particle.

Suppose that the particle has velocity $\boldsymbol{v}$ relative to the inertial coordinate system t, x, y, z. Then $\mathrm{d}t/\mathrm{d}\tau = \gamma(v)$, where $v = |\boldsymbol{v}|$, and

$$\frac{\mathrm{d}x}{\mathrm{d}\tau} = \frac{\mathrm{d}t}{\mathrm{d}\tau}\frac{\mathrm{d}x}{\mathrm{d}t} = \gamma(v)v_1. \tag{6.6}$$

Similarly $\mathrm{d}y/\mathrm{d}\tau = \gamma(v)v_2$, $\mathrm{d}z/\mathrm{d}\tau = \gamma(v)v_3$, where v_1, v_2, v_3 are the components of $\boldsymbol{v}$. Hence the components of the four-velocity V are

$$(V^0, V^1, V^2, V^3) = \gamma(v)(c, v_1, v_2, v_3)\,. \tag{6.7}$$

In other words V decomposes into temporal and spatial parts as

$$V = \gamma(v)(c, \boldsymbol{v})\,. \tag{6.8}$$

If we denote by X the displacement four-vector from the origin of the inertial coordinate system, then we have

$$\frac{\mathrm{d}X}{\mathrm{d}\tau} = V\,.$$

Since the particle is moving in a straight line at constant speed, the components of V are constant. We can therefore integrate to obtain

$$X = \tau V + K$$

where K is a constant four-vector. It follows that if A and B are events on the worldline with B happening after A, then the displacement four-vector from A to B is τV, where τ is the proper time from A to B.

Proposition 6.4

If V is the four-velocity of a particle, then $g(V, V) = c^2$.

Proof

We shall give two proofs. First, in an inertial coordinate system in which the particle is at rest, $V^0 = c$, $V^1 = V^2 = V^3 = 0$, so

$$g(V, V) = c^2\,. \tag{6.9}$$

But $g(V, V)$ is an invariant. Therefore (6.9) is valid in any inertial coordinate system.

Second, in a general inertial coordinate system

$$\begin{aligned} g(V, V) &= (V^0)^2 - (V^1)^2 - (V^2)^2 - (V^3)^2 \\ &= \gamma(v)^2(c^2 - \boldsymbol{v}\,.\,\boldsymbol{v}) \\ &= c^2\,. \end{aligned}$$

□

Thus the four-velocity is a timelike four-vector. It is also future-pointing since $V^0 = c\gamma(v) > 0$. It is a useful tool for solving problems of relative motion without making coordinate transformations overtly, which can be a clumsy way to proceed.

Problem 6.5 (Addition of Velocities)

An observer has velocity $\boldsymbol{u}$ and a particle has velocity $\boldsymbol{v}$ in some inertial coordinate system. Find the speed w of the particle relative to the observer in terms of $\boldsymbol{u}$ and $\boldsymbol{v}$.

Solution

We have to find the speed of the particle in an inertial coordinate system in which the observer is at rest. Let U be the four-velocity of the observer and let V be the four-velocity of the particle. In the given inertial coordinate system,

$$U = \gamma(u)(c, \boldsymbol{u}) \quad \text{and} \quad V = \gamma(v)(c, \boldsymbol{v}) .$$

Hence

$$g(U, V) = \gamma(u)\gamma(v)(c^2 - \boldsymbol{u} \,.\, \boldsymbol{v}) .$$

Now consider an inertial coordinate system in which the observer is at rest. In this system

$$U = (c, 0), \qquad V = \gamma(w)(c, \boldsymbol{w}) ,$$

where $\boldsymbol{w} \,.\, \boldsymbol{w} = w^2$. Hence

$$g(U, V) = \gamma(w) .$$

But $g(U, V)$ is invariant. Therefore

$$c^2\gamma(w) = \gamma(u)\gamma(v)(c^2 - \boldsymbol{u} \,.\, \boldsymbol{v}) .$$

On putting $\gamma(w) = 1/\sqrt{1 - w^2/c^2}$ and solving for w, one finds that

$$w = \frac{c\sqrt{c^2(\boldsymbol{u} - \boldsymbol{v}) \,.\, (\boldsymbol{u} - \boldsymbol{v}) - u^2v^2 + (\boldsymbol{u} \,.\, \boldsymbol{v})^2}}{c^2 - \boldsymbol{u} \,.\, \boldsymbol{v}}$$

which reduces to the classical formula $w = |\boldsymbol{u} - \boldsymbol{v}|$ when $u, v \ll c$. □

Problem 6.6

A non-accelerating observer O has four-velocity U. Let A and B be two events. Show that O reckons that A and B are simultaneous if and only if the displacement vector X from A to B is orthogonal to U; that is, if and only if $g(U, X) = 0$.

Solution

Pick an inertial coordinate system in which O is at rest. Then U and X have components

$$U = (c, 0, 0, 0), \qquad X = (X^0, X^1, X^2, X^3),$$

where X^0 is the time separation of A and B, multiplied by c. Thus in this inertial coordinate system

$$g(U, X) = cX^0 .$$

Now O reckons that A and B are simultaneous if and only if $X^0 = 0$; that is if and only if $g(U, X) = 0$. □

Problem 6.7

Two observers O and O' are travelling in straight lines at constant speeds. Show that there is a pair of events A and A', with A on the worldline of O and A' on the worldline of O', which O and O' agree are simultaneous.

Solution

Let the four-velocities of O and O' be V and V', respectively. Pick two events B (on the worldline of O) and B' (on the worldline of O'). Then if A is any other event on the worldline of O, the displacement vector from B to A is τV for some $\tau \in \mathbb{R}$; and if A' is any other event on the worldline of O', the displacement vector from B' to A' is $\tau' V'$ for some $\tau' \in \mathbb{R}$. The displacement vector from A to A' is therefore

$$Y = X - \tau V + \tau' V'$$

where X is the displacement vector from B to B'.

From the result of the preceding problem, it is sufficient to find A and A' such that both $g(Y, V) = 0$ and $g(Y, V') = 0$. That is, we must solve

$$\begin{aligned} g(X, V) - c^2\tau + g(V', V)\tau' &= 0 \\ g(X, V') - g(V, V')\tau + c^2\tau' &= 0 \end{aligned} \qquad (6.10)$$

for τ and τ'. From above, $g(V, V') = c^2\gamma(w)$, where w is the speed of O' relative to O. So either $w \neq 0$, in which case $g(V, V') > c^2$ and Equations (6.10) have a unique solution for τ and τ', or $w = 0$ and $V = V'$, in which case there are infinitely many solutions. □

Problem 6.8 (Lorentz Contraction)

A rod is of length L_0 in its rest frame. In a second inertial frame, it is oriented in the direction of the unit vector $\boldsymbol{e}$ and is moving with velocity $\boldsymbol{v}$. Show that its length in the second frame is

$$L = \frac{L_0\sqrt{c^2 - v^2}}{\sqrt{c^2 - v^2 \sin^2\theta}},$$

where θ is the angle between $\boldsymbol{e}$ and $\boldsymbol{v}$.

Solution

Let V be the four-velocity of the rod. Let A be an event at one end of the rod, and let B and C be events at the other end such that B is simultaneous with A in the frame in which the rod is moving and C is simultaneous with A in the rest frame of the rod. The problem is to find the distance L between A and B, measured in the frame in which the rod is moving.

Let X be the displacement four-vector from A to B and let Y be the displacement four-vector from A to C. Then

$$Y = X + \tau V$$

for some scalar τ and

$$L^2 = -g(X, X), \qquad L_0^2 = -g(Y, Y).$$

Also $g(V, Y) = 0$ since A and C are simultaneous in the rest frame of the rod.

We find τ by calculating $g(X, V)$ in two different ways. In the frame in which the rod is moving, we have $X = (0, L\boldsymbol{e})$ and $V = \gamma(v)(c, \boldsymbol{v})$, so

$$g(X, V) = -L\gamma(v)\boldsymbol{e}\,.\,\boldsymbol{v} = -Lv\gamma(v)\cos\theta\,.$$

We also have

$$g(X, V) = g(Y - \tau V, V) = -\tau c^2\,.$$

Hence $\tau = Lv\gamma(v)\cos\theta/c^2$ and therefore

$$\begin{aligned} L^2 &= -g(X, X) \\ &= -g(Y - \tau V, Y - \tau V) \\ &= L_0^2 - \tau^2 c^2 \\ &= L_0^2 - c^{-2}L^2\gamma(v)^2 v^2 \cos^2\theta\,. \end{aligned}$$

The result follows after a little algebra. □

6.4 Four-acceleration

So far, we have investigated only motion with constant velocity. We now consider the general motion of an accelerating particle. Its worldline is a curve in space–time given by

$$x = x(t), \quad y = y(t), \quad z = z(t),$$

where t, x, y, z are inertial coordinates. Let E at time t and E' at time $t + \delta t$ be two nearby events on the worldline. We define the *proper time* from E to E' to be the time $\delta\tau$ measured in a second inertial coordinate system in which the particle is instantaneously at rest at the event E.

In the t, x, y, z coordinate system, the displacement four-vector X from E to E' is

$$X = (c, \boldsymbol{v})\, \delta t$$

where $\boldsymbol{v}$ is the velocity of the particle. From the discussion in §5.11,

$$c^2 \delta\tau^2 = g(X, X) = (c^2 - \boldsymbol{v} \,.\, \boldsymbol{v})\, \delta t^2 \,. \tag{6.11}$$

As in the case of a non-accelerating particle, therefore,

$$\frac{\mathrm{d}\tau}{\mathrm{d}t} = \frac{1}{\gamma(v)} \tag{6.12}$$

where v is the speed of the particle.

Definition 6.9

The parameter τ defined up to the addition of a constant by Equation (6.12) is called the *proper time* along the particle worldline.

Suppose that the particle carries a clock with a mechanism that is unaffected by the acceleration (a pendulum clock, for example, would not do). Such an 'ideal' clock could be expected to show the same time interval between two nearby events on the worldline as that measured in an inertial frame in which the particle is instantaneously at rest. In other words, it could be expected to show the proper time τ. We assume that this is in fact the case.

Clock Hypothesis

Proper time is the time measured by an accelerating ideal clock travelling with the particle.

Example 6.10

Suppose that the worldline is given in inertial coordinates by

$$x = R\cos(\omega t), \qquad y = R\sin(\omega t), \qquad z = 0 \tag{6.13}$$

with R and ω constant. That is, the particle is moving in a circle of radius R with angular speed ω. Then $v = R\omega$ and

$$\frac{\mathrm{d}\tau}{\mathrm{d}t} = \sqrt{1 - R^2\omega^2/c^2}\,. \tag{6.14}$$

Up to an additive constant, therefore,

$$\tau = t\sqrt{1 - R^2\omega^2/c^2}\,.$$

We must have $R^2\omega^2 < c^2$ for the motion to be physically possible (otherwise the particle would be moving faster than light).

Here the speed is constant and τ is a constant multiple of coordinate time. More generally, if the first two equations in (6.13) are replaced by $x = R\cos\theta$, $y = R\sin\theta$, where θ is some general function of t, then (6.14) still holds, with $\omega = \mathrm{d}\theta/\mathrm{d}t$; but τ is no longer a constant multiple of t.

The four-velocity V is defined in the same way as for uniform motion, with the components given by

$$V^0 = c\frac{\mathrm{d}t}{\mathrm{d}\tau}, \quad V^1 = \frac{\mathrm{d}x}{\mathrm{d}\tau}, \quad V^2 = \frac{\mathrm{d}y}{\mathrm{d}\tau}, \quad V^3 = \frac{\mathrm{d}z}{\mathrm{d}\tau}\,.$$

The proof of the four-vector transformation rule is the same as in the uniform case—it uses only the constancy of the Lorentz transformation matrix between the two coordinate systems; and V has the same decomposition

$$V = \gamma(v)(c, \boldsymbol{v})$$

into temporal and spatial parts, with

$$\boldsymbol{v} = \left(\frac{\mathrm{d}x}{\mathrm{d}t}, \frac{\mathrm{d}x}{\mathrm{d}t}, \frac{\mathrm{d}z}{\mathrm{d}t}\right).$$

The difference is that V now depends on τ. Thus it makes sense to define a *four-acceleration* A by

$$A^a = \frac{\mathrm{d}V^a}{\mathrm{d}\tau}\,. \tag{6.15}$$

The four-vector transformation rule for the components of A follows by differentiating (6.5) with respect to τ, and again using the fact that L is constant.

Example 6.11

If the worldline is given by Equation (6.13), then the four-acceleration has components

$$-\gamma(v)^2 R\omega^2 (0, \cos(\omega t), \sin(\omega t), 0),$$

with $v = R\omega$.

In a general inertial coordinate system, A has spatial and temporal parts

$$\begin{aligned} A &= \gamma(v)\frac{\mathrm{d}}{\mathrm{d}t}\Big(\gamma(v)(c, \boldsymbol{v})\Big) \\ &= \frac{v\gamma(v)^4}{c^2}\frac{\mathrm{d}v}{\mathrm{d}t}(c, \boldsymbol{v}) + \gamma(v)^2\left(0, \frac{\mathrm{d}\boldsymbol{v}}{\mathrm{d}t}\right). \end{aligned} \tag{6.16}$$

So in an inertial coordinate system in which the particle is instantaneously rest,

$$V = (c, 0), \qquad A = (0, \boldsymbol{a})$$

where $\boldsymbol{a} = \mathrm{d}^{\boldsymbol{v}}/\mathrm{d}t$ is the ordinary acceleration. It follows that

$$g(A, V) = 0, \quad g(V, V) = c^2, \quad g(A, A) = -a^2,$$

where a is the magnitude of the acceleration measured in the instantaneous rest frame; that is, the acceleration 'felt' by an observer moving with the particle.

Example 6.12

For the worldline given by (6.13),

$$g(A, A) = -\gamma(v)^4 R^2\omega^4 = -\frac{R^2\omega^4}{(1 - R^2\omega^2/c^2)^2}, \tag{6.17}$$

and therefore

$$a = \frac{R\omega^2}{1 - R^2\omega^2/c^2}.$$

So the acceleration a 'felt' by a someone travelling in a circle with the particle is larger than the classical value of $a = R\omega^2$ by a factor $1/(1 - R^2\omega^2/c^2)$. As ω approaches the limiting value of c/R, we have $a \to \infty$.

6.5 Constant Acceleration

Suppose that $y = z = 0$ along the worldline in some fixed inertial coordinate system, and that a is constant. The components of V and A in the fixed coordinate system are

$$(c\dot{t}, \dot{x}, 0, 0) \quad \text{and} \quad (c\ddot{t}, \ddot{x}, 0, 0),$$

where the dot denotes $\mathrm{d}/\mathrm{d}\tau$, differentiation with respect to proper time. Hence

$$c^2\dot{t}^2 - \dot{x}^2 = c^2 \quad \text{and} \quad c^2\ddot{t}^2 - \ddot{x}^2 = -a^2 .$$

By differentiating the first equation and substituting into the second,

$$c\ddot{t} = a\sqrt{\dot{t}^2 - 1}, \qquad \ddot{x} = a\dot{t} .$$

Hence, with a suitable choice of origin for τ,

$$\dot{t} = \cosh(a\tau/c) \quad \text{and} \quad \dot{x} = c\sinh(a\tau/c)$$

and, with a suitable choice of origin for t and x,

$$t = \frac{c}{a}\sinh\left(\frac{a\tau}{c}\right) \quad \text{and} \quad x = \frac{c^2}{a}\cosh\left(\frac{a\tau}{c}\right) .$$

The worldline is a hyperbola with asymptotes $ct = \pm x$ (Figure 6.1). Although at each event the acceleration measured in an inertial coordinate system in which the particle is instantaneously at rest is always the same, the velocity of the particle never exceeds c.

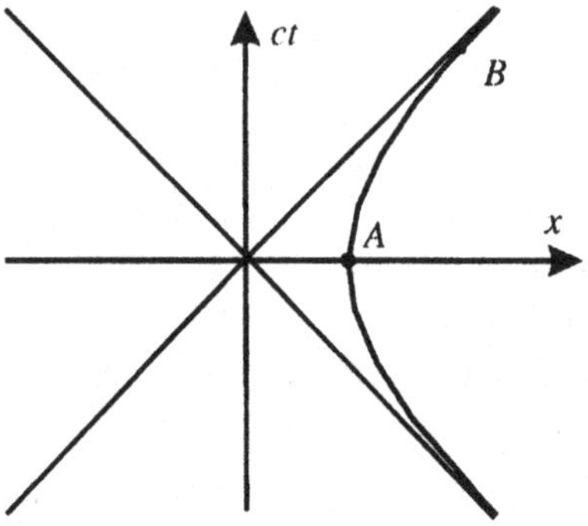

Figure 6.1 A constant acceleration worldline

Consider two events on the worldline:

A at which $x = c^2/a$ and $t = 0$ and
B at which $x = (c^2/a)\cosh(a\tau/c)$ and $t = (c/a)\sinh(a\tau/c)$.

The proper time between them measured by a clock travelling with the particle is τ. Measured in the fixed inertial coordinate system t, x, y, z, the time between them is

$$t = \frac{c}{a}\sinh\left(\frac{a\tau}{c}\right)$$

and the distance between them is

$$x = \frac{c^2}{a}\left(\cosh\left(\frac{a\tau}{c}\right) - 1\right).$$

If $a = g$ (the acceleration due to gravity) and $\tau = 10$ years, then

$$t = 11{,}000 \text{ years} \quad \text{and} \quad x = 11{,}000 \text{ light-years.}$$

One could exploit this result for inter-stellar travel [8]. Suppose that a spaceship sets out from earth to travel to a star 22,000 light-years from earth. It could do this by accelerating for 10 years with $a = g$ and then decelerating for 10 years with $a = g$, arriving at the star after a passage of 20 years, measured on the spaceship. It could then return in the same way, arriving back at earth after the passage of a further 20 years, again measured on the spaceship. The catch is that the time between departure and return measured on earth is 44,000 years (the interest on the loan that financed the expedition would by then have exceeded the GNP of the entire universe).

Two points should be noted. First the fact that the spaceship can complete a round trip of 44,000 light-years in 40 years does not involve a violation of the prohibition on faster-than-light travel since the distance and the time are measured in different frames; the distance is measured on earth and the time is measured on the spaceship. The calculation illustrates an important fact; it is consistent with special relativity to travel as far as you like in as short a time as you like, *provided that* the distance is measured before you set out and the time is measured along your worldline.

Second, the fact that the time between departure and return measured on earth is not the same as the time measured on the spaceship does not violate the principle of relativity. There is an asymmetry between the two; the spaceship is accelerating, but the earth is not. It is sometimes said incorrectly that there is a paradox here (the 'twin paradox', since it is stated in terms of two twins, one in the spaceship, the other on earth). But the result is only paradoxical if one forgets that although uniform motion has only a relative meaning in Einstein's special theory, acceleration is absolute.

Example 6.13

Two rockets moving with constant acceleration have respective worldlines

$$t = \frac{c}{a}\sinh\left(\frac{a\tau}{c}\right), \qquad x = \frac{c^2}{a}\cosh\left(\frac{a\tau}{c}\right)$$

and

$$t = \frac{c}{a'}\sinh\left(\frac{a'\tau'}{c}\right), \qquad x = \frac{c^2}{a'}\cosh\left(\frac{a'\tau'}{c}\right)$$

in some inertial coordinate system, where a and a' are the respective accelerations, and τ and τ' are the proper time parameters along the respective worldlines.

Consider the displacement four-vector X from the event A at proper time τ on the first worldline to the event A' at proper time $\tau' = a'\tau/a$ on the second. That is

$$X = \frac{c^2(a - a')}{aa'}\Big(\sinh(a\tau/c), \cosh(a\tau/c), 0, 0\Big)$$

in the inertial coordinate system. Then we have

$$g(X, X) = -\frac{c^4(a - a')^2}{a^2a'^2}, \qquad g(X, V) = 0, \qquad g(X, V') = 0\,,$$

where V is the four-velocity of the first rocket at A, and V' is the four-velocity of the second at A'. Therefore observers on both rockets reckon that A and A' are simultaneous, and separated by a distance

$$c^2\left|\frac{1}{a'} - \frac{1}{a}\right|,$$

which is independent of τ. So the two observers reckon that the two rockets remain a constant distance apart.

A continuous family of such worldlines with varying values of a is a model for an accelerating rigid rod; the worldlines are those of the individual particles making up the rod. It is rigid because the distance between two neighbouring particles is unchanging in the instantaneous rest frame of either of them. Of course in the coordinates t, x, y, z, the particles get further apart as τ increases through positive values; but the increase in separation is exactly matched by the Lorentz contraction.

6.6 Continuous Distributions

We shall look now not just at the motion of a single particle, but a continuous distribution of particles. We suppose that in an inertial coordinate system, there are σ particles per unit volume, moving with velocity $\boldsymbol{v}$, where σ and $\boldsymbol{v}$ are differentiable functions of position and time. What is the number of particles per unit volume in a second inertial coordinate system? We can answer this by using the following.

Proposition 6.14

Under change of inertial coordinates, $c\sigma$ and $\sigma\boldsymbol{v}$ transform as the temporal and spatial parts of a four-vector.

Proof

We know that $\gamma(v)(c, \boldsymbol{v})$ transforms as a four-vector (the particle four-velocity). So the task is to show that $\sigma/\gamma(v)$ is the same in all inertial coordinate systems.

Suppose first that $\boldsymbol{v}$ is constant, so that all the particles are moving in straight lines with the same velocity. Let $\tilde{t}$, $\tilde{x}$, $\tilde{y}$,$\tilde{z}$ be an inertial coordinate system in which the particles are rest, and suppose that there are $\tilde{\sigma}$ particles per unit volume in these coordinates. Consider the particles which at $\tilde{t} = 0$ occupy a small cube, four of whose vertices are

$$A : (0,0,0), \quad B : (L,0,0), \quad C : (0,L,0), \quad D : (0,0,L) \tag{6.18}$$

in the coordinate system $\tilde{x}, \tilde{y}, \tilde{z}$. Let t, x, y, z be a second inertial coordinate system related to $\tilde{t}, \tilde{x}, \tilde{y}, \tilde{z}$ by the standard Lorentz transformation with velocity v. In the these coordinates, A, B, C, D have worldlines

$$\begin{aligned} A &: \; x = vt, \quad y = z = 0 \\ B &: \; x = vt + L\sqrt{1 - v^2/c^2}, \quad y = z = 0 \\ C &: \; x = vt, \quad y = L, \quad z = 0 \\ D &: \; x = vt, \quad y = 0, \quad z = L, \end{aligned} \tag{6.19}$$

where we have used the result of §4.8 in dealing with B. Therefore the cube appears in the second inertial coordinate system to be a cuboid of volume $L^3/\gamma(v)$ moving with velocity $\boldsymbol{v} = v\boldsymbol{i}$; so there are $\sigma = \gamma(v)\tilde{\sigma}$ particles per unit volume in the second inertial coordinate system. Therefore $\sigma/\gamma(v)$ is independent of $\boldsymbol{v}$. Since σ and $\gamma(v)$ are also unchanged under rotation, it follows that $\sigma/\gamma(v)$ is the same in every inertial coordinate system.

When $\boldsymbol{v}$ and σ are general functions of position and time, then we can apply the same argument after first restricting attention to a small neighbourhood of an event in which $\boldsymbol{v}$ and σ can be treated as approximately constant. □

If particles are neither created nor destroyed, then the argument in §2.12 shows that σ and $\boldsymbol{v}$ must satisfy the *continuity equation*

$$\operatorname{div}(\sigma \boldsymbol{v}) + \frac{\partial \sigma}{\partial t} = 0 .$$

That is,

$$\operatorname{Div} N = 0 ,$$

where N is the four-vector $(c\sigma, \sigma\boldsymbol{v})$.

6.7 *Rigid Body Motion

A special case is that in which the particles make up a *rigid body*. That is when a (non-inertial) observer travelling with any one of the particles reckons that the distances to nearby particles remain constant. In classical physics, a rigid body has six degrees of freedom, three for the position of some given point of the body, and three for rotations about this point. Under the action of appropriate forces, the three position coordinates and the three angular coordinates determining its orientation can be any prescribed functions of time. In special relativity, this is not true; the rigidity condition on its own severely restricts the possible motions, independently of any dynamical considerations. This is one reason why we did not base the operational definition of distance on the notion of a rigid measuring rod.

Proposition 6.15

The motion of a distribution of particles with four-velocity field V is rigid if and only if $g(X, \nabla_X V) = 0$ at every event for every four-vector X orthogonal to V.

Proof

Let t, x, y, z be an inertial coordinate system in which one of the particles is instantaneously at rest at the event E at which all the coordinates are zero. If the motion is rigid, then the nearby particles must be moving as if they were part of a classical rigid body, since for slow motion the distances to the nearby particles will be just as in Galilean relativity. At $t = 0$, the velocities of particles near the origin will be given by

$$\boldsymbol{v} = \boldsymbol{\omega} \wedge \boldsymbol{r} \tag{6.20}$$

for some vector $\boldsymbol{\omega}$, where $\boldsymbol{r} = (x, y, z)$, to the first order in $\boldsymbol{r}$. So if $\boldsymbol{x}$ is a three-vector, then

$$(\boldsymbol{x} \,.\, \boldsymbol{\nabla})\boldsymbol{v} = \boldsymbol{\omega} \wedge \boldsymbol{x}\,, \tag{6.21}$$

is orthogonal to $\boldsymbol{x}$ at E.

Let V denote the four-velocity field of the particles. Then in the inertial coordinate system

$$V = \gamma(v)(c, \boldsymbol{v})$$

where $\boldsymbol{v}$ is a function of the coordinates. At E, we have $\boldsymbol{v} = 0$, $\gamma(v) = 1$, and the first partial derivatives of γ vanish, since, for example,

$$\frac{\partial}{\partial x}\left(\frac{1}{\sqrt{1 - v^2/c^2}}\right) = 0$$

when $v = 0$.

A four-vector X orthogonal to V at E is of the form $(0, \boldsymbol{x})$. If X is such a four-vector, then at E we have

$$g(X, \nabla_X V) = -\boldsymbol{x} \,.\, \big((\boldsymbol{x} \,.\, \boldsymbol{\nabla})\boldsymbol{v}\big) = 0\,.$$

But the left-hand side is an invariant, so we conclude that for rigid motion $g(X, \nabla_X V) = 0$ everywhere for every X orthogonal to V.

To prove the converse, it is easy to work back to (6.21), and then to (6.20) by expanding the components of $\boldsymbol{v}$ in a Taylor series, to the first order in $\boldsymbol{r}$. □

The condition $g(X, \nabla_X V) = 0$ whenever $g(X, V) = 0$ is called the *Born rigidity condition* on the four-velocity V. If it is satisfied by a four-vector field V (not necessarily normalized), then it is also satisfied by $K = fV$ for any nonzero function f on space–time, since if $g(X, K) = 0$ then $g(X, V) = 0$ and

$$g(X, \nabla_X K) = fg(X, \nabla_X V) + g(X, \operatorname{Grad} f)g(X, V) = 0\,.$$

The converse is also true; if we are given any future-pointing timelike vector field K such that $g(X, \nabla_X K) = 0$ for every X orthogonal to K, then we get a possible rigid motion by putting $V = cK/\sqrt{g(K, K)}$, so that V satisfies the four-velocity normalization condition $g(V, V) = c^2$.

There are two types of rigid motion.

- The first is *planar motion*, for which $\omega = 0$ throughout the body. In this case the four-velocity field V is determined by the worldline of one particle of the body. If the four-velocity is V_E at an event E on the worldline of the particle, then $V = V_E$ at every event which is simultaneous with E in the instantaneous rest-frame of the particle at E. The whole body can accelerate, but it cannot rotate.
- The second is *isometric motion*, in which V is proportional to a four-vector field K satisfying the *Killing vector condition*

$$g(X, \nabla_X K) = 0 \qquad \text{for every four-vector } X.$$

Such four-vector fields are determined by ten parameters; in a fixed inertial frame, they are of the form $K = (\kappa, \boldsymbol{k})$, where

$$\kappa = \boldsymbol{a} \cdot \boldsymbol{r} + \beta, \qquad \boldsymbol{k} = ct\boldsymbol{a} + \boldsymbol{\Omega} \wedge \boldsymbol{r} + \boldsymbol{b}, \tag{6.22}$$

where $\boldsymbol{r} = (x, y, z)$, and $\boldsymbol{a}, \boldsymbol{b}, \boldsymbol{\Omega}$ and β are constant. In this case the body rotates (if $\boldsymbol{\Omega} \neq 0$), but not in an arbitrary way; the rotation at later times is determined by the rigidity condition and by the initial motion, independently of any dynamical considerations. The ten parameters are β and the nine components of $\boldsymbol{a}, \boldsymbol{b}, \boldsymbol{\Omega}$; for a given choice of these, the body can only occupy the region of space–time in which K is timelike.

Exercise 6.1

Show that if the temporal and spatial parts of K are given by (6.22), then $g(X, \nabla_X K) = 0$ for every four-vector X.

The Herglotz–Noether theorem states these are the only possible rigid motions (a good account can be found in [12]). The proof of the theorem is beyond the scope of this book, but is an interesting exercise for anyone familiar with differential geometry. It begins with the restatement of the rigidity condition in terms of the projection map from Minkowski space onto the space M of integral curves of the four-vector field V. The condition is equivalent to the existence of a set of orthonormal 1-forms θ_0, θ_1, θ_2, θ_3 such that $\theta_0(V) = c$, and such that θ_1, θ_2, and θ_3 are pull-backs of forms on M. One uses the Cartan calculus to obtain the curvature 2-form, and works from the condition that the curvature of the metric on Minkowski space should vanish.

6.8 Visual Observation

The language of special relativity can sometimes mislead. For example, the statement 'a measuring rod appears to an observer moving in the direction of the rod to have contracted' is true only if the phrase 'appears ... to an observer' is interpreted in terms of a particular measuring procedure; the observer must set up inertial coordinates and then determine the distance between the worldlines of the two ends of the rod. It is tempting to make the erroneous assumption that other classically equivalent measurements will give the same result—for example, a measurement of the angle subtended by the two ends of the rod at known distance. In fact it does not because it involves *visual observation*; the motion of the observer also affects the measured angle between the trajectories of the photons arriving at the observer from the two ends of the rod. In fact a moving rod does not even 'appear' to be straight when observed visually.

To understand what an inertial observer actually sees at a particular event E on his worldline, we must consider the photon worldlines that pass through E. These are the generators of the light-cone of E. An event A is on the past light-cone of E if the displacement four-vector K from A to E is null and future-pointing. In the observer's inertial frame, it has temporal and spatial parts

$$K = (\kappa, \boldsymbol{k}),$$

where $\kappa = \sqrt{\boldsymbol{k} . \boldsymbol{k}} > 0$. Light emitted at A arrives at the observer from the direction of the unit vector $-\boldsymbol{k}/\kappa$.

Suppose that B is a second event on the past light-cone of E and that the displacement vector from B to E is the null four-vector $L = (\lambda, \boldsymbol{\ell})$. If the angle between the directions from which light from A and B arrives at the observer is θ, then

$$\cos\theta = \frac{\boldsymbol{k} . \boldsymbol{\ell}}{\kappa\lambda} . \tag{6.23}$$

The following proposition re-expresses this formula in terms of invariants.

Proposition 6.16

Suppose that A and B are events on the past light-cone of an event E on the worldline of an observer with four-velocity V. Let K and L be the respective displacement four-vectors from A to E and from B to E. Then the observer measures the angle between light rays reaching him at E from A and B to be θ, where

$$\cos\theta = 1 - \frac{c^2 g(K, L)}{g(K, V) g(L, V)} .$$

Proof

In the observer's frame, $V = (c, 0)$. Therefore

$$g(V,K) = c\kappa, \qquad g(V,L) = c\lambda, \qquad g(K,L) = \kappa\lambda - \boldsymbol{k}\,.\,\boldsymbol{\ell}\,.$$

It follows that

$$1 - \frac{c^2 g(K,L)}{g(K,V)g(L,V)} = \frac{\boldsymbol{k}\,.\,\boldsymbol{\ell}}{\kappa\lambda} = \cos\theta\,.$$

□

Problem 6.17 (Stellar Aberration)

An observer measures the angle subtended by two distant stars to be θ. Show that a second observer moving relative to the first with speed v directly away from one of the stars measures the angle to be θ', where

$$\cos\theta' = \frac{c\cos\theta - v}{c - v\cos\theta}\,.$$

What happens as $c \to \infty$?

The formula in the problem is the *stellar aberration* formula.

Solution

We take A and B to be events at which light from the stars is emitted, and we suppose that the second observer moves relative to the first directly away from the star at A. Then we have

$$g(K,L) = \kappa\lambda(1 - \cos\theta)\,,$$

where $K = (\kappa, \boldsymbol{k})$, $L = (\lambda, \boldsymbol{\ell})$ in the first observer's frame, and the second observer has velocity $v\boldsymbol{k}/\kappa$ relative to the first.

In the first observer's frame, the four-velocity of the second observer is given by

$$V = \gamma(v)(c, v\boldsymbol{k}/\kappa)\,.$$

Hence,

$$g(K,V) = \gamma(v)\left(c\kappa - \frac{v\boldsymbol{k}\,.\,\boldsymbol{k}}{\kappa}\right) = \kappa\gamma(v)(c - v)$$

and

$$g(L,V) = \gamma(v)\left(c\lambda - \frac{v\boldsymbol{k}\,.\,\boldsymbol{\ell}}{\kappa}\right) = \lambda\gamma(v)(c - v\cos\theta)\,.$$

From Proposition 6.16, therefore,

$$\begin{aligned}\cos\theta' &= 1-\frac{c^2 g(K,L)}{g(K,V)g(L,V)}\\ &= 1-\frac{c^2(1-\cos\theta)}{\gamma(v)^2(c-v)(c-v\cos\theta)}\\ &= 1-\frac{(c+v)(1-\cos\theta)}{c-v\cos\theta}\\ &= \frac{c\cos\theta-v}{c-v\cos\theta}\,.\end{aligned}$$

If $\theta \neq 0$, so that the stars are separated in the sky, then $\cos\theta' \to -1$ as $v \to c$. So as the second observer looks in the direction of his motion relative to the first and accelerates towards the velocity of light, all the stars appear to move across the sky to positions directly ahead. This includes the stars that were initially behind him, but not directly behind him. □

A striking example of the distinction between visual observation and coordinate measurement is provided by a moving sphere; whatever the speed of the sphere, the visually observed outline is always circular, despite the fact that, according to coordinate measurements, it is squashed along the direction of its motion by the Lorentz contraction.[1]

To see this, consider the light rays reaching an observer at the origin from the visually observed outline of a stationary sphere. If the three-vector from the centre of the sphere to the observer is $\boldsymbol{x}$ and if the sphere subtends an angle 2α at the observer, then light reaching the observer from the outline of the sphere will travel in the direction of one of the unit vectors $\boldsymbol{e}$ that satisfies

$$\boldsymbol{e}\,.\,\boldsymbol{x} = |\boldsymbol{x}|\cos\alpha\,. \tag{6.24}$$

Consider a photon emitted from the sphere at the event A that reaches the observer at the event E at which $t = x = y = z = 0$. Suppose that the displacement four-vector K from A to E has temporal and spatial parts $(\kappa, \boldsymbol{k})$ in the observer's frame. Then $\kappa = \sqrt{\boldsymbol{k}\,.\,\boldsymbol{k}}$ since K is future-pointing and null. If the photon appears to the observer to come from the outline of the sphere, then $\boldsymbol{e} = \boldsymbol{k}/\kappa$ satisfies (6.24). So we can characterize the 'outline events' E by the condition

$$\kappa|\boldsymbol{x}|\cos\alpha - \boldsymbol{k}\,.\,\boldsymbol{x} = 0\,.$$

That is, $g(K,X) = 0$, where X is the spacelike four-vector with temporal and spatial parts $(|\boldsymbol{x}|\cos\alpha, \boldsymbol{x})$ in the observer's frame.

[1] This was first pointed out, surprisingly late in the development of relativity, by Roger Penrose in 1959 [6].

In the frame of another inertial observer, $X = (\xi', \boldsymbol{x}')$ and $K = (\kappa', \boldsymbol{k}')$, with $\kappa' = |\boldsymbol{k}'|$. Since $g(K, X)$ is invariant, we shall have for the 'outline events',

$$\kappa'\xi' - \boldsymbol{k} \,.\, \boldsymbol{x}' = 0 ,$$

and hence the photons reaching the second observer from the visually observed outline of the sphere travel in the directions $\boldsymbol{e} = \boldsymbol{k}'/\kappa'$ that satisfy

$$\boldsymbol{e} \,.\, \boldsymbol{x}' = |\boldsymbol{x}'| \cos\alpha' ,$$

where $\cos\alpha' = \xi'/|\boldsymbol{x}'|$. Therefore they appear to the second observer to come from a sphere with centre in the direction of $-\boldsymbol{x}'$, with outline subtending an angle $2\alpha'$. To a moving observer, therefore, a sphere still has the visual appearance of a sphere, but of a different size. It still has a circular outline, and does not 'look' squashed.

Exercise 6.2

Show that

$$c \cot\alpha' = g(X, V)/\sqrt{-g(X, X)} ,$$

where V is the four-velocity of the second observer, and hence that

$$\cot\alpha' = \gamma(v)\left(\cot\alpha - \frac{\boldsymbol{x} \,.\, \boldsymbol{v} \operatorname{cosec}\alpha}{c|\boldsymbol{x}|}\right) ,$$

where $\boldsymbol{v}$ is the velocity of the second observer relative to the first.

Note that if the second observer moves directly away from the sphere with velocity v, then

$$\cot\alpha' = \gamma(v)\left(\cot\alpha - \frac{v}{c}\operatorname{cosec}\alpha\right) ,$$

so $\alpha' \to \pi$ as $v \to c$. So as the observer accelerates away from the sphere, the outline grows until it fills the whole sky, apart from a small hole directly ahead.

EXERCISES

6.3. Person P travels from event A to event B with constant four-velocity U in proper time σ, and then from event B to event C with constant four-velocity $V \neq U$ in proper time τ. A second person P' travels from A to C with constant four-velocity V' in proper time τ'. Show that

$$\tau' V' = \sigma U + \tau V ,$$

and hence that

$$\tau'^2 = \sigma^2 + \tau^2 + 2\gamma(v)\sigma\tau ,$$

where v is the speed of P in the second part of his journey relative to a frame in which he is at rest in the first part of his journey. Show also that

$$\tau' > \sigma + \tau .$$

What is the corresponding relationship between σ, τ, τ' in classical physics? [See Exercise 5.8.]

6.4. †A particle travels in a straight line relative to an inertial coordinate system. Show that

$$g(A, A) = -c^2 \left(\frac{\mathrm{d}\phi}{\mathrm{d}\tau}\right)^2 ,$$

where A is the four-acceleration, τ is proper time, and ϕ is the rapidity.

6.5. +Two rockets accelerating along the x-axis in opposite directions with constant acceleration a have worldlines

$$x = -\frac{c^2\cosh(a\tau/c)}{a}, \qquad t = \frac{c\sinh(a\tau/c)}{a},$$

and

$$x = \frac{c^2\cosh(a\tau/c)}{a}, \qquad t = \frac{c\sinh(a\tau/c)}{a},$$

respectively.

Let $Z(\tau)$ denote the displacement four-vector from the event A at proper time $-\tau$ on the first worldline to the event B at proper time τ on the second worldline. Show that

(i) $g(Z, Z)$ is independent of τ;

(ii) Z is orthogonal to the four-velocity of the first rocket at A and to the four-velocity of the second rocket at B.

Deduce that observers in the two rockets reckon that A and B are simultaneous for every choice of τ, and that they both think that the distance between A and B is independent of τ.

Thus the two rockets are always the same distance apart, according to the observers. Explain this apparent absurdity.

6.6. +Four distant stars S_i are observed. Let θ_{ij} denote the angle between the directions to S_i and S_j. Show that

$$\frac{(1-\cos\theta_{12})(1-\cos\theta_{34})}{(1-\cos\theta_{13})(1-\cos\theta_{24})}$$

is independent of the motion of the observer.

7
Relativistic Collisions

In Einstein's theory, the inertial coordinate systems that observers use to label events in Minkowski space are related by Lorentz transformations. This gives a consistent, if unfamiliar, framework within which to understand relative motion. It is one in which the conspicuous incompatibility between Maxwell's equations and the principle of relativity is resolved because the velocity of photons is always the same, whatever the motion of the source or the observer. But there is a price; Newton's laws are invariant under Galiliean transformations, but not under Lorentz transformations. In fact, Galilean invariance is a corollary of Newton's laws. So if Einstein's picture of space–time is the correct one, then we must revise the basic principles of dynamics.

The problem is already evident in collisions (interactions at a point). In Newtonian theory, when two particles of equal mass collide and coalesce, the velocity of the resulting particle is the mean of the velocities of the particles before the collision. If the particles are moving along the x-axis with velocities u_1 and u_2 before the collision, then afterwards the combined particle has velocity

$$u_3 = \tfrac{1}{2}(u_1 + u_2)\,. \tag{7.1}$$

But this is incompatible with the Lorentz transformation. An observer moving with velocity v along the x-axis reckons that the velocities of the particles are

$$u_i' = \frac{u_i - v}{1 - u_i v/c^2}\,,$$

($i = 1, 2, 3$). Since this is nonlinear in u_i, Equation (7.1) cannot hold for the transformed velocities for all v. So the Newtonian laws of conservation of mass and momentum are incompatible with special relativity.

As elsewhere, the way forward is to let go of a classical idea that seems so self-evident that it hardly needs to be identified as a basic element of Newtonian mechanics. The resolution comes not by changing conservation of momentum, but by re-examining the notion of mass. We have to set aside the naive identification of 'mass' with 'quantity of matter', and reconsider the apparently self-evident law that mass is conserved. As with other basic quantities, such as distance and time, mass must be given an operational definition; it must be defined in terms of the procedure by which it is measured.

Mass enters Newtonian mechanics in two ways, as *inertial mass*, the constant m in the second law $F = ma$, and as *gravitational mass*, the constants m and m' in the inverse-square law $F = Gmm'/r^2$. It is the latter that one measures by weighing a body. But this is not a good starting point for our operational definition since there is no sensible way to include gravitational interactions in special relativity; in relativity, gravity appears ultimately not as a force, but as the curvature of a space–time that looks like Minkowski space only over small distances and short times.

Instead we begin with a direct *dynamical* measurement of mass in collisions. In point interactions, we can avoid having to face up to the awkward conflict between action at a distance and the axiom of Einstein's theory that no influence can be transmitted from one particle to another at a speed faster than that of light.

7.1 The Operational Definition of Mass

Our starting point is the fact that the Newtonian conservation laws (§1.6) hold to a high degree of accuracy in collisions in which the velocities of the particles are much less than the velocity of light. Given a standard mass M, therefore, an observer can assign a mass m to any other particle by colliding it at low speed with the standard mass, measuring the resulting velocities, and applying the Newtonian law of conservation of momentum. Since this becomes exact as the velocities go to zero, the observer can in principle use a limiting procedure to measure m when the particle is at rest.

Definition 7.1

The *rest mass* of a particle is the mass measured by low-speed collisions in an inertial coordinate system in which the particle is at rest.

Rest mass is an intrinsic quantity associated with a particle.

7.2 Conservation of Four-momentum

Each particle in a collision has a rest mass m (a scalar) and a four-velocity V (a four-vector). The four-vector $P = mV$ is called the *four-momentum* of the particle. It has temporal and spatial parts

$$P = \big(m\gamma(v)c, m\gamma(v)\boldsymbol{v}\big),$$

where $\boldsymbol{v}$ is the three-velocity. As $\boldsymbol{v} \to 0$, $\gamma(v) = 1 + O(v^2/c^2)$ and

$$P = (mc, m\boldsymbol{v}) + O(v^2/c^2).$$

So if all the velocities are so small that terms in v^2/c^2 can be neglected, then the Newtonian laws of conservation of mass and momentum are equivalent to the conservation of the temporal and spatial parts of four-momentum.

We need to replace the Newtonian laws by statements that are equivalent when v^2/c^2 can be neglected, but which are otherwise compatible with Lorentz transformations. A very straightforward possibility is to adopt the hypothesis that four-momentum is always conserved.

Four-Momentum Hypothesis

If the incoming particles in a collision have four-momenta $P_1, P_2, \ldots, P_k$ and the outgoing particles have four-momenta $P_{k+1}, P_{k+2}, \ldots, P_n$, then

$$\sum_1^k P_i = \sum_{k+1}^n P_i . \tag{7.2}$$

The justification for this is, first, that it is equivalent to the Newtonian laws of conservation of mass and momentum for low speed collisions and, second, since it is a relationship between four-vectors, it is compatible with Lorentz transformations; if it holds in one inertial frame, then it holds without approximation in every inertial frame.

Whatever the velocities of the particles, we can still take the temporal and spatial parts of (7.2) to obtain

$$\begin{aligned} \sum_1^k m_i\gamma(v_i) &= \sum_{k+1}^n m_i\gamma(v_i) \\ \sum_1^k m_i\gamma(v_i)\boldsymbol{v}_i &= \sum_{k+1}^n m_i\gamma(v_i)\boldsymbol{v}_i , \end{aligned}$$

where the m_i are the rest masses of the particles. These take the same form as the Newtonian laws of mass and momentum conservation when we identify $m\gamma(v)$ with inertial mass and $m\gamma(v)\boldsymbol{v}$ with three-momentum.

Definition 7.2

Suppose that a particle of rest mass m has velocity $\boldsymbol{v}$ relative to some inertial coordinate system. The quantities $m_I = m\gamma(v)$ and $\boldsymbol{p} = m_I\boldsymbol{v}$ are the *inertial mass* and the *three-momentum* of the particle relative to the inertial coordinate system.

Four-momentum conservation is equivalent to conservation of inertial mass and of three-momentum (in every inertial coordinate system). The new feature of the relativistic theory is that the inertial mass of a particle increases with its velocity, albeit only very slightly for velocities much less than that of light.

Rest mass is a scalar—by its operational definition, it is an intrinsic quantity—but inertial mass is different in different inertial coordinate systems. Rest mass and inertial mass are equal for a particle at rest.

Problem 7.3

A particle of rest mass M is at rest when it splits into two particles, each of rest mass m, which move with velocities $(u, 0, 0)$ and $(-u, 0, 0)$. Show that $M = 2m\gamma(u)$.

Solution

By conservation of four-momentum

$$M(c, 0, 0, 0) = m\gamma(u)(c, u, 0, 0) + m\gamma(u)(c, -u, 0, 0)\,. \tag{7.3}$$

Hence $M = 2m\gamma(u)$. Note that $M > 2m$; and that $m \to 0$ as $u \to c$, for fixed M. □

7.3 Equivalence of Mass and Energy

In a general collision, it is not rest mass that is conserved, but the temporal part

$$P^0 = m\gamma(v)$$

of the four-momentum. Now

$$\gamma(v) = 1 + \frac{v^2}{2c^2} + O\left(\frac{v^4}{c^4}\right)\,.$$

So if we neglect terms of order v^4/c^4, but keep terms of order v^2/c^2, then

$$P^0 = \frac{1}{c}\left(mc^2 + \tfrac{1}{2}mv^2\right),$$

where m is the rest mass. Thus cP^0 is the sum of the Newtonian kinetic energy and a much larger term mc^2, which also has the dimensions of energy.

Definition 7.4

For a particle of rest mass m, the quantity mc^2 is the *rest energy* of the particle.

Any collision that involves a gain or loss of kinetic energy, such as an explosion or an inelastic collision, must involve a corresponding loss or gain in the total rest energies of the particles; kinetic energy can be traded for rest mass, and vice versa.

When v is comparable with c, we must include the higher order terms in v/c. We then have

$$cP^0 = m_I c^2,$$

where $m_I = \gamma(v)m$ is the inertial mass. We call cP^0 the *total energy* of the particle relative to the inertial coordinate system. It is usually denoted by E, so that

$$E = m_I c^2.$$

Total energy is conserved in collisions.

The total energy of a particle is frame-dependent; it is different in different coordinate systems and is at its minimum in an inertial coordinate system in which the particle is at rest, where it is equal to the rest energy. So the energy of a particle at rest is given by the famous formula

$$E = mc^2.$$

In classical physics, the energy that can be stored in a given mass in the form of heat or chemical or nuclear energy is, in principle, unlimited. In special relativity, by contrast, mass and energy are equivalent, apart from the factor c^2, and any stored energy in a particle contributes to its mass. If a body is heated, for example, then its rest mass increases, usually by a negligible amount. Conversely the maximum energy that can be extracted from a stationary body of rest mass m is its rest energy mc^2. The upper limit is very large. Even in the explosion of an atomic bomb, only about 0.1% of the rest mass is released as other forms of energy.

Problem 7.5 (Elastic collision)

An elastic collision is one in which the rest masses of the particles are unchanged, so there is no exchange between kinetic and 'stored' energy.

A particle of rest mass m is moving with velocity $\boldsymbol{u}$ relative to some inertial coordinate system when it collides elastically with a second particle, also of rest mass m, which is at rest. After the collision, the particles have velocities $\boldsymbol{v}$ and $\boldsymbol{w}$. Show that if θ is the angle between $\boldsymbol{v}$ and $\boldsymbol{w}$, then

$$\cos\theta = \frac{c^2}{vw}\left(1 - \sqrt{1 - v^2/c^2}\right)\left(1 - \sqrt{1 - w^2/c^2}\right). \tag{7.4}$$

Solution

By conservation of four-momentum

$$m(c, 0) + m\gamma(u)(c, \boldsymbol{u}) = m\gamma(v)(c, \boldsymbol{v}) + m\gamma(w)(c, \boldsymbol{w}).$$

Therefore

$$1 + \gamma(u) = \gamma(v) + \gamma(w) \tag{7.5}$$

$$\gamma(u)\boldsymbol{u} = \gamma(v)\boldsymbol{v} + \gamma(w)\boldsymbol{w}. \tag{7.6}$$

By taking the modulus-squared of both sides of (7.6),

$$\gamma(u)^2u^2 = \gamma(v)^2v^2 + \gamma(w)^2w^2 + 2\gamma(v)\gamma(w)\boldsymbol{v}\,.\,\boldsymbol{w}.$$

But $\gamma(u)^2u^2 = c^2(\gamma(u)^2 - 1)$ (a very useful identity). Hence

$$2\gamma(v)\gamma(w)\boldsymbol{v}\,.\,\boldsymbol{w} = c^2\big(\gamma(u)^2 - \gamma(v)^2 - \gamma(w)^2 + 1\big).$$

But by squaring both sides of (7.5),

$$1 + 2\gamma(u) + \gamma(u)^2 = \gamma(v)^2 + \gamma(w)^2 + 2\gamma(v)\gamma(w),$$

from which we obtain

$$\begin{aligned} 1 + \gamma(u)^2 - \gamma(v)^2 - \gamma(w)^2 &= 2\gamma(v)\gamma(w) - 2\gamma(u) \\ &= 2\big(\gamma(v)\gamma(w) + 1 - \gamma(v) - \gamma(w)\big) \\ &= 2(\gamma(v) - 1)(\gamma(w) - 1). \end{aligned}$$

Hence

$$\frac{\boldsymbol{v}\,.\,\boldsymbol{w}}{vw} = \frac{c^2\big(\gamma(v) - 1\big)\big(\gamma(w) - 1\big)}{vw\gamma(v)\gamma(w)},$$

from which (7.4) follows.

Note that $\cos\theta > 0$ whenever $v, w > 0$, so the angle between $\boldsymbol{v}$ and $\boldsymbol{w}$ is always acute. Also $\cos\theta \to 1$, so $\theta \to 0$ as $v, w \to c$. This is in contrast with the Newtonian theory, in which kinetic energy is conserved as well as momentum, since the collision is elastic. So the Newtonian theory gives

$$\boldsymbol{u} = \boldsymbol{v} + \boldsymbol{w} \qquad \text{and} \qquad u^2 = v^2 + w^2 ,$$

from which it follows that $\boldsymbol{v} \,.\, \boldsymbol{w} = 0$ and hence that $\theta = \pi/2$ for all values of $\boldsymbol{v}$ and $\boldsymbol{w}$.

The fact that θ can be seen in particle tracks in high-energy accelerator collisions to be very much less than $\pi/2$ offers clear confirmation of the relativistic collision laws. □

EXERCISES

7.1. A particle of rest mass M and total energy E collides with a particle of rest mass m at rest. Show that the sum E' of the total energies of the two particles in the frame in which their centre of mass is at rest is given by

$$E'^2 = (M^2 + m^2)c^4 + 2Emc^2 .$$

[The centre of mass is defined in such a way that its four-velocity V is proportional to the total four-momentum of the two particles.]

7.2. It is intended to produce a J/ψ particle of rest mass M by colliding a positron of rest mass $m \ll M$ at velocity $\boldsymbol{v}$ with a stationary electron, also of rest mass m. Show that for this to be possible the energy $mc^2(\gamma(v) - 1)$ of the positron above its rest energy must be at least the threshold energy E_{min}, defined by

$$E_{\text{min}} = Mc^2\left(\frac{M}{2m} - \frac{2m}{M}\right) .$$

Explain why it would be more efficient to attempt this experiment in an accelerator in which electron and positron beams collide with equal and opposite velocities.

7.3. $^{+}$The rocket problem.

(i) Relative to a given inertial frame, a rocket is travelling along the x-axis with velocity u and has rest mass m when it ejects a particle of rest mass M and velocity $-v$. Afterwards, the rocket has rest mass m' and velocity u'. Relative to the rest frame of the rocket before the ejection, the ejected particle has velocity $-w$. Show that

$$m'\gamma(u')(u-u') + M\gamma(v)(u+v) = 0$$

and

$$m'^2 = m^2 + M^2 - 2mM\gamma(w)\,.$$

(ii) Show that

$$\gamma(v)\begin{pmatrix} c \\ -v \end{pmatrix} = \gamma(u)\gamma(w)\begin{pmatrix} 1 & u/c \\ u/c & 1 \end{pmatrix}\begin{pmatrix} c \\ -w \end{pmatrix}.$$

(iii) Deduce that

$$mm'\gamma(u')\gamma(u)(u'-u) = mM\gamma(w)w = \tfrac{1}{2}w(m^2 - m'^2 + M^2)\,.$$

(iv) Now suppose that the rocket accelerates along the x-axis of the given frame by firing out a stream of particles in the negative x direction. As it does so, its rest mass decreases. Show by putting $m' = m + \delta m$, $u' = u + \delta u$ in (iii), and by taking the limit as $M \to 0$, that

$$m\frac{\mathrm{d}u}{\mathrm{d}m} + w(1 - u^2/c^2) = 0\,,$$

where u is the speed of the rocket relative to the inertial coordinate system, m is the rest mass, and w is the speed of the particles relative to the rocket.

(v) Now suppose that w and the acceleration are constant (as measured in the instantaneous rest frame of the rocket at each event). Show that

$$\frac{m_0}{m} = \left(\frac{c+u}{c-u}\right)^{c/2w} = \mathrm{e}^{a\tau/w}\,,$$

where m_0 is the initial rest mass and τ is the proper time along the rocket's worldline. Estimate the total amount of fuel required for the expedition described in §6.5 to a star 22,000 light-years away, and back, at acceleration g, on the assumptions that $w =$ 1000 mph and that the only filling station is on earth. Comment on the economic viability of chemical rockets for travel across galactic distances.

8
Relativistic Electrodynamics

8.1 Lorentz Transformations of E and B

The requirement that Maxwell's equations should be consistent with the principle of relativity implies that the velocity of photons must be independent of the motion of their source and of the observer. That, in conjunction with other plausible assumptions, leads to the conclusion that inertial coordinate systems must be related by Lorentz transformations. It is not immediately obvious, however, that this chain of reasoning is reversible, and that Maxwell's equations are in fact invariant. To show that, we must find the transformation rule for the components of the electric and magnetic fields. We must address the question: if an observer moves with velocity $\boldsymbol{v}$ through a given electromagnetic field, what electric and magnetic fields will he observe and do the observed fields satisfy Maxwell's equations?

We answer the first part of the question by considering observations of the motion of a charged particle relative to an inertial frame. The path of a particle moving slowly through an electric field $\boldsymbol{E}$ and magnetic field $\boldsymbol{B}$ is determined by the Lorentz force law. If the particle has velocity $\boldsymbol{v}$, momentum $\boldsymbol{p}$ and charge e, then

$$\frac{\mathrm{d}\boldsymbol{p}}{\mathrm{d}t} = e(\boldsymbol{E} + \boldsymbol{v} \wedge \boldsymbol{B})\,. \tag{8.1}$$

By measuring the trajectory for different velocities, an observer can in principle determine $\boldsymbol{E}$ and $\boldsymbol{B}$ at each point. Knowing how to apply an arbitrary Lorentz transformation to $\boldsymbol{E}$ and $\boldsymbol{B}$ is equivalent to knowing how to extend the equation of motion (8.1) to any $\boldsymbol{v}$ with $|\boldsymbol{v}| < c$. For if we know how to do the former, then we can transform to a frame in which the particle is moving slowly, find its trajectory, and then transform back to the original coordinates.

The transformation law for $\boldsymbol{E}$ and $\boldsymbol{B}$ must correctly encode the behaviour of particles moving at high speed through electric and magnetic fields. This has two features (amply verified in particle accelerators). First, the rest mass m and charge e of a particle are unchanged by interaction with the fields; and, second, if we take $\boldsymbol{p}$ to be the spatial part of the four-momentum, then Equation (8.1) holds for any $\boldsymbol{v}$ with $|\boldsymbol{v}| < c$. That is, the motion of a charged particle at any velocity is governed by

$$\frac{\mathrm{d}\boldsymbol{p}}{\mathrm{d}t} = e(\boldsymbol{E} + \boldsymbol{v} \wedge \boldsymbol{B}) \qquad \text{where} \qquad \boldsymbol{p} = m\gamma(v)\boldsymbol{v}\,. \tag{8.2}$$

From this starting point, we derive the following.

Proposition 8.1

The four-acceleration $A = (\alpha, \boldsymbol{a})$ of a particle of rest mass m and charge e moving with velocity $\boldsymbol{v}$ in an electromagnetic field is given by

$$mc\alpha = e\gamma(v)\boldsymbol{E}\,.\,\boldsymbol{v}, \qquad m\boldsymbol{a} = e\gamma(v)(\boldsymbol{E} + \boldsymbol{v} \wedge \boldsymbol{B})\,.$$

Proof

Since m is constant and $\boldsymbol{p} = m\gamma(v)\boldsymbol{v}$, the second equation follows from (8.2), together with

$$\boldsymbol{a} = \gamma(v)\frac{\mathrm{d}}{\mathrm{d}t}\Big(\gamma(v)\boldsymbol{v}\Big)\,.$$

The first follows from the orthogonality of the four-acceleration $A = (\alpha, \boldsymbol{a})$ and the four-velocity $V = \gamma(v)(c, \boldsymbol{v})$, which implies that $c\alpha = \boldsymbol{a}\,.\,\boldsymbol{v}$. □

Let F denote the 4×4 matrix

$$F = \begin{pmatrix} 0 & E_1 & E_2 & E_3 \\ -E_1 & 0 & -cB_3 & cB_2 \\ -E_2 & cB_3 & 0 & -cB_1 \\ -E_3 & -cB_2 & cB_1 & 0 \end{pmatrix}. \tag{8.3}$$

This is called the *electromagnetic field*. The entries in F are denoted by F_{ab}, $a, b = 0, 1, 2, 3$; they are the *components* of the electromagnetic field. An observer measures the electromagnetic field at an event by measuring the four-acceleration of charged particles, and by using Proposition 8.1.

Proposition 8.2 (Tensor Property of F)

Suppose that the inertial coordinate systems of two inertial observers are related by (5.10). Then the electromagnetic fields measured by the two observers at an event are related by $F' = L^{\mathrm{t}}FL$.

In the language of Chapter 9, $F = \big(F_{ab}\big)$ is a *covariant tensor*, of type $(0, 2)$.

Proof

We shall write the four-velocity and four-acceleration of a charged particle as column vectors V and A. Then Proposition 8.1 gives

$$cmA = e\gamma(v)\begin{pmatrix} v_1E_1 + v_2E_2 + v_3E_3 \\ cE_1 + v_2cB_3 - v_3cB_2 \\ cE_2 + v_3cB_1 - v_1cB_3 \\ cE_3 + v_1cB_2 - v_2cB_1 \end{pmatrix}.$$

But we also have

$$egFV = e\gamma(v)\begin{pmatrix} 0 & E_1 & E_2 & E_3 \\ E_1 & 0 & cB_3 & -cB_2 \\ E_2 & -cB_3 & 0 & cB_1 \\ E_3 & cB_2 & -cB_1 & 0 \end{pmatrix}\begin{pmatrix} c \\ v_1 \\ v_2 \\ v_3 \end{pmatrix},$$

where, as usual, g is the diagonal matrix with diagonal entries $1, -1, -1, -1$. The left-hand sides are equal. In matrix notation, therefore, the equation of motion is

$$cmA = egFV\,.$$

But four-velocity and four-acceleration transform as four-vectors, so $V = LV'$ and $A = LA'$; and by the pseudo-orthogonality property, we have $L^{-1} = gL^{\mathrm{t}}g$. Hence

$$cmLA' = egFLV'$$

and therefore

$$cmA' = egL^{\mathrm{t}}FLV'$$

since g^2 is the identity. But in the second coordinate system,

$$cmA' = eF'gV'\,.$$

Since both equations for A' hold whatever the four-velocity, we conclude that $F' = L^{\mathrm{t}}F'L$. □

Note that the transformation preserves the skew-symmetry of F. It should be remarked also that the following assumption is implicit in this proof.

Invariance of Charge

The charge of a particle is the same in all inertial coordinate systems.

One piece of physical evidence for this is the overall neutrality of matter. When at rest, electrons and protons have equal and opposite charges. In an atom the electrons are moving much faster than the protons in the nucleus. If the charge of a particle depended on its velocity, then there could not be an exact balance between the electric charges of the electrons and the protons.

8.2 The Four-Current and the Four-potential

We now turn to the invariance of Maxwell's equations. We can write them in terms of the potentials in the Lorenz gauge as

$$\Box\phi = \frac{\rho}{\epsilon_0}, \qquad \Box c\boldsymbol{A} = \frac{\boldsymbol{J}}{c\epsilon_0}, \tag{8.4}$$

with

$$\frac{1}{c^2}\frac{\partial\phi}{\partial t} + \operatorname{div}(\boldsymbol{A}) = 0\,. \tag{8.5}$$

We shall look first at the transformation of the right-hand sides of (8.4). Suppose that we have a continuous distribution of charged particles, with σ particles per unit volume, each moving with velocity $\boldsymbol{v}$, where σ and $\boldsymbol{v}$ are differentiable functions of the inertial coordinates. Then by Proposition 6.14, $c\sigma$ and $\sigma\boldsymbol{v}$ are the temporal and spatial parts of a four-vector. If each particle has charge e, then

$$c\rho = e\sigma c, \qquad \boldsymbol{J} = e\sigma\boldsymbol{v}\,.$$

Under the assumption that e is invariant, therefore, $(c\rho, \boldsymbol{J})$ transforms as a four-vector. The same must be true of any distribution of charges since the four-vector transformation rule is linear and since we can sum the contributions to ρ and $\boldsymbol{J}$ from different populations of particles with different charges moving with different velocities.

Definition 8.3

The *current four-vector* J is the four-vector with temporal part $c\rho$ and spatial part $\boldsymbol{J}$.

Note that continuity equation (2.26) now takes the invariant form

$$\operatorname{Div} J = 0 .$$

The left-hand sides of (8.4) are dealt by the following proposition, which states in effect that ϕ and $c\boldsymbol{A}$ fit together to form a four-vector. It is phrased more awkwardly because the potentials are not uniquely determined by F.

Proposition 8.4

Let ϕ and $\boldsymbol{A}$ be scalar and vector potentials in the Lorenz gauge for an electromagnetic field in some inertial coordinate system. Suppose that ϕ and $c\boldsymbol{A}$ are transformed as the temporal and spatial parts of a four vector under some inhomogeneous Lorentz transformation. Then the results are scalar and vector potentials for the transformed field, again in the Lorenz gauge.

Proof

Note first that if we transform $(\phi, c\boldsymbol{A})$ as the temporal and spatial parts of a four-vector Φ, then the Lorenz gauge condition is preserved since it then takes the invariant form

$$\operatorname{Div} \Phi = 0 .$$

Given ϕ and $\boldsymbol{A}$, define a matrix-valued function M by

$$M = \frac{1}{c} \begin{pmatrix} \partial_t \\ c\partial_x \\ c\partial_y \\ c\partial_z \end{pmatrix} \Big(\phi,\ -cA_1,\ -cA_2,\ -cA_3\Big) .$$

Then $F = M - M^{\mathrm{t}}$ since, for example, the second entry in the first row of $M - M^{\mathrm{t}}$ is

$$-\frac{\partial A_1}{\partial t} - \frac{\partial \phi}{\partial x} = E_1 ,$$

and the third entry in the second row is

$$-c\frac{\partial A_2}{\partial x} + c\frac{\partial A_1}{\partial y} = -c\big(\operatorname{curl} \boldsymbol{A}\big)_3 = -cB_3 .$$

The other entries are dealt with in a similar way.

Under the transformation (5.10), we have

$$\Big(\partial_{t'},\ c\partial_{x'},\ c\partial_{y'},\ c\partial_{z'}\Big) = \Big(\partial_t,\ c\partial_x,\ c\partial_y,\ c\partial_z\Big) L$$

(see §5.12). Define ϕ' and $\boldsymbol{A}'$ in the coordinate system $t', x', y'z'$ by applying the four-vector transformation rule to ϕ and $\boldsymbol{A}$. That is,

$$\begin{pmatrix} \phi' \\ cA'_1 \\ cA'_2 \\ cA'_3 \end{pmatrix} = L^{-1} \begin{pmatrix} \phi \\ cA_1 \\ cA_2 \\ cA_3 \end{pmatrix} .$$

Equivalently, since $L^{-1} = gL^{\mathrm{t}}g$,

$$\left(\phi',\ -cA'_1,\ c - A'_2,\ -cA'_3\right) = \left(\phi,\ -cA_1,\ -cA_2,\ -cA_3\right)L .$$

Put

$$M' = \frac{1}{c} \begin{pmatrix} \partial_{t'} \\ c\partial_{x'} \\ c\partial_{y'} \\ c\partial_{z'} \end{pmatrix} \left(\phi',\ -cA'_1,\ -cA'_2,\ -cA'_3\right) .$$

Then we have $M' = L^{\mathrm{t}}ML$ and hence $M' - M'^{\mathrm{t}} = F'$. It follows that ϕ' and $\boldsymbol{A}'$ are potentials for the electric and magnetic fields in the primed coordinates. □

Definition 8.5

The four-vector Φ with temporal and spatial parts ϕ and $c\boldsymbol{A}$ is called the *four-potential.*

In terms of the four-potential in the Lorenz gauge, Maxwell's equations take the explicitly invariant form

$$\Box\Phi = \frac{1}{c\epsilon_0}J, \qquad \mathrm{Div}\,\Phi = 0 .$$

Hence we have the following proposition.

Proposition 8.6

If Maxwell's equations hold for the electromagnetic field F, then they also hold for the transformed field F'.

Under gauge transformations, $\Phi \mapsto \Phi + \mathrm{Grad}\,u$, with the Lorenz condition preserved if $\Box u = 0$.

8.3 Transformations of $\boldsymbol{E}$ and $\boldsymbol{B}$

The tensor property of F allows us to relate the components of the electric and magnetic fields in different inertial frames. If the inertial coordinate systems are related by a rotation, then, in block form,

$$L = \begin{pmatrix} 1 & 0 \\ 0 & H \end{pmatrix},$$

where $H \in \mathrm{SO}(3)$. We then have

$$\begin{pmatrix} E_1' \\ E_2' \\ E_3' \end{pmatrix} = H^{\mathrm{t}} \begin{pmatrix} E_1 \\ E_2 \\ E_3 \end{pmatrix} \quad \text{or equivalently} \quad \begin{pmatrix} E_1 \\ E_2 \\ E_3 \end{pmatrix} = H \begin{pmatrix} E_1' \\ E_2' \\ E_3' \end{pmatrix}$$

which is the normal transformation rule for the components of a three-vector, and

$$\begin{pmatrix} 0 & -B_3' & B_2' \\ B_3' & 0 & -B_1' \\ -B_2' & B_1' & 0 \end{pmatrix} = H^{\mathrm{t}} \begin{pmatrix} 0 & -B_3 & B_2 \\ B_3 & 0 & -B_1 \\ -B_2 & B_1 & 0 \end{pmatrix} H.$$

However, this is equivalent to

$$\begin{pmatrix} B_1 \\ B_2 \\ B_3 \end{pmatrix} = H \begin{pmatrix} B_1' \\ B_2' \\ B_3' \end{pmatrix},$$

as can be checked directly by taking H to be, in turn,

$$\begin{pmatrix} \cos\theta & \sin\theta & 0 \\ -\sin\theta & \cos\theta & 0 \\ 0 & 0 & 1 \end{pmatrix}, \quad \begin{pmatrix} \cos\theta & 0 & \sin\theta \\ 0 & 1 & 0 \\ -\sin\theta & 0 & \cos\theta \end{pmatrix}, \quad \begin{pmatrix} 1 & 0 & 0 \\ 0 & \cos\theta & \sin\theta \\ 0 & -\sin\theta & \cos\theta \end{pmatrix}.$$

So the components of $\boldsymbol{B}$ also transform as the components of a three-vector—although this is *not* true if H is orthogonal, but has negative determinant, and so reverses the 'handedness' of the spatial axes.

In the case of the standard Lorentz transformation,

$$L = \begin{pmatrix} \gamma & \gamma u/c & 0 & 0 \\ \gamma u/c & \gamma & 0 & 0 \\ 0 & 0 & 1 & 0 \\ 0 & 0 & 0 & 1 \end{pmatrix},$$

where $\gamma = \gamma(u)$, the tensor property $F' = L^{\mathrm{t}} F L$ gives

$$E_1' = E_1, \quad E_2' = \gamma(E_2 - uB_3), \quad E_3' = \gamma(E_3 + uB_2) \tag{8.6}$$

and

$$B_1' = B_1, \quad B_2' = \gamma(B_2 + uE_3/c^2), \quad B_3' = \gamma(B_3 - uE_2/c^2) \tag{8.7}$$

so the transformation mixes electric and magnetic fields. If $u \ll c$, then

$$\boldsymbol{E}' = \boldsymbol{E} + \boldsymbol{u} \wedge \boldsymbol{B},$$

where $\boldsymbol{u} = u\boldsymbol{i}$. Thus an observer moving slowly with velocity $\boldsymbol{u}$ through a pure magnetic field $\boldsymbol{B}$ with $\boldsymbol{E} = 0$ sees an electric field $\boldsymbol{u} \wedge \boldsymbol{B}$, as we anticipated in Chapter 2.

It follows from (8.6) and (8.7) that

$$\begin{aligned} \boldsymbol{E}' . \boldsymbol{B}' &= E_1 B_1 + \gamma^2 (E_2 - uB_3)(B_2 + uE_3/c^2) \\ &\quad + \gamma^2 (E_3 + uB_2)(B_3 - uE_2/c^2) \\ &= \boldsymbol{E} . \boldsymbol{B} . \end{aligned} \tag{8.8}$$

Hence $\boldsymbol{E} . \boldsymbol{B}$ is invariant under standard Lorentz transformations. Clearly it is also invariant under rotations. Hence it is an invariant of the electromagnetic field—it is the same in every inertial coordinate system. Another invariant is $\boldsymbol{E} . \boldsymbol{E} - c^2 \boldsymbol{B} . \boldsymbol{B}$.

It follows, for example, that if $\boldsymbol{E}$ and $\boldsymbol{B}$ are not orthogonal in some inertial coordinate system, then $\boldsymbol{E} \neq 0$ and $\boldsymbol{B} \neq 0$ in every inertial coordinate system.

Example 8.7 (The Field of a Uniformly Moving Charge)

Suppose that the electromagnetic field is generated by a charge e which is stationary at the origin in the coordinate system t', x', y', z'. Then $\boldsymbol{B}' = 0$ and

$$E_1' = \frac{kx'}{r'^3}, \quad E_2' = \frac{ky'}{r'^3}, \quad E_3' = \frac{kz'}{r'^3}, \tag{8.9}$$

where $k = e/4\pi\epsilon_0$ and $r'^2 = x'^2 + y'^2 + z'^2$.

What is the field in the t, x, y, z coordinates, in which the particle is moving with velocity $\boldsymbol{u} = u\boldsymbol{i}$? We answer this by applying the inverse of (8.6) and (8.7), which is given by replacing u by $-u$ and interchanging the primed and unprimed field components. From the standard Lorentz transformation,

$$x' = \gamma(x - ut), \quad y' = y, \quad z' = z. \tag{8.10}$$

In the t, x, y, z coordinates, therefore,

$$E_1 = \frac{k\gamma(x - ut)}{r'^3}, \quad E_2 = \frac{k\gamma y}{r'^3}, \quad E_3 = \frac{k\gamma z}{r'^3}, \tag{8.11}$$

and

$$B_1 = 0, \quad B_2 = -\frac{k\gamma u z}{c^2 r'^3}, \quad B_3 = \frac{k\gamma u y}{c^2 r'^3}. \tag{8.12}$$

In vector form

$$\boldsymbol{E} = \frac{\gamma k \boldsymbol{R}}{r'^3}, \qquad \boldsymbol{B} = \frac{1}{c^2} \boldsymbol{u} \wedge \boldsymbol{E} \tag{8.13}$$

where $\boldsymbol{R}$ is the position vector from $(ut, 0, 0)$, the position of the charge at time t, to (x, y, z), the point at which the field is measured.

It remains to express r' in terms of t, x, y, z coordinates. By substituting from (8.10) into the definition of r', we have

$$\begin{aligned} r'^2 &= \gamma^2(x-ut)^2 + y^2 + z^2 \\ &= (x-ut)^2 + y^2 + z^2 + \frac{u^2(x-ut)^2}{c^2-u^2} \\ &= \left(1 + \frac{u^2\cos^2\theta}{c^2-u^2}\right) \boldsymbol{R}\,.\,\boldsymbol{R} \\ &= \gamma^2(1 - u^2\sin^2\theta/c^2)\,\boldsymbol{R}\,.\,\boldsymbol{R} \end{aligned} \tag{8.14}$$

where θ is the angle between $\boldsymbol{R}$ and $\boldsymbol{u}$. Finally, therefore, we deduce that $\boldsymbol{E}$ is in the direction of $\boldsymbol{R}$ and has magnitude

$$E = \frac{e}{4\pi\epsilon_0 R^2}\left[\frac{1}{\gamma^2(1-u^2\sin^2\theta/c^2)^{3/2}}\right]. \tag{8.15}$$

When $u \ll c$, the factor in square brackets is one, $\boldsymbol{E}$ is the inverse-square-law electrostatic field generated by a charge at the point $(ut, 0, 0)$ and $\boldsymbol{B}$ is the magnetic field of a slow-moving charge, as given in Chapter 2.

The electric field of a fast moving-charge is reduced from the familiar electrostatic field by a factor $1/\gamma^2$ on $\theta = 0$ (the trajectory of the charge) and enhanced by a factor γ on $\theta = \pi/2$, the plane through the charge orthogonal to its velocity. In the limit $u \to c$, the field is entirely concentrated on this plane.

8.4 Linearly Polarized Plane Waves

Consider the equations for the four-potential in the Lorenz gauge in the absence of sources,

$$\Box\Phi = 0\,, \quad \mathrm{Div}\,\Phi = 0\,. \tag{8.16}$$

Let us look for a solution of the form $\Phi = P\cos(\Omega + \epsilon)$, where P is a constant four-vector, ϵ is a constant, and $\Omega \neq 0$ is a linear function of the coordinates. That is

$$\Omega = \omega(t - \boldsymbol{e}\,.\,\boldsymbol{r}/c)$$

for some constants ω and $\boldsymbol{e}$. Without loss of generality, we can take ω to be positive.

Put $K = c\,\mathrm{Grad}\,\Omega$. Then K is a constant four-vector with temporal part ω and Equations (8.16) are equivalent to

$$g(K,K)P\cos(\Omega+\epsilon) = 0, \quad g(K,P)P\sin(\Omega+\epsilon) = 0\,.$$

So for Φ to be a solution, K must be a future-pointing null vector, and P must be a spacelike four-vector orthogonal to K. In this case, each component of Φ is a solution of the wave equation with frequency four-vector K.

If we decompose P and K by putting

$$K = \omega(1, \boldsymbol{e}), \qquad P = (0, \boldsymbol{p}) + \beta\omega(1, \boldsymbol{e})$$

for some $\beta \in \mathbb{R}$ (chosen so that $P - \beta K$ has vanishing temporal part), then

$$\boldsymbol{E} = \frac{\omega}{c}\boldsymbol{p}\sin(\Omega + \epsilon), \qquad \boldsymbol{B} = \frac{1}{c}\boldsymbol{e} \wedge \boldsymbol{E},$$

so the corresponding solution of Maxwell's equation is a linearly polarized monochromatic plane wave with frequency ω.

Definition 8.8

The four-vector $K = c\,\mathrm{Grad}\,\Omega$ is the *frequency four-vector* of the linearly polarized electromagnetic wave with four-potential $\Phi = P\cos(\Omega + \epsilon)$.

A general monochromatic plane wave can be generated by a four-potential of the form $\Phi = P\cos\Omega + Q\sin\Omega$ where P and Q are constant four-vectors. In this case also we define the frequency four-vector to be $K = c\,\mathrm{Grad}\,\Omega$.

In the inertial coordinate system of an observer with four-velocity U, we have

$$K = \omega(1, \boldsymbol{e}), \qquad U = (c, 0)\,.$$

Therefore the frequency measured by the observer is

$$\omega = \frac{g(K, U)}{c}\,.$$

In some other inertial coordinate system,

$$K = \omega'(1, \boldsymbol{e}') \qquad \text{and} \qquad U = \gamma(u)(c, \boldsymbol{u}) \tag{8.17}$$

where ω' and $\boldsymbol{e}'$ are the frequency and direction of propagation of the wave in the new inertial coordinate system. By evaluating the right-hand side of (8.17) in the new system,

$$\omega = \omega'\gamma(u)\left(1 - \frac{\boldsymbol{e}' . \boldsymbol{u}}{c}\right).$$

This is the relativistic *Doppler formula*; it gives the frequency measured by an observer who is moving with velocity $\boldsymbol{u}$ through a plane wave of frequency ω' propagating in the direction of the unit vector $\boldsymbol{e}'$. In contrast to the classical theory, there is a Doppler shift even in the case $\boldsymbol{e}'.\boldsymbol{u} = 0$, in which the motion is orthogonal to the line of sight.

8.5 Electromagnetic Energy

Consider a cloud of charged particles, each of rest mass m and charge e. Each particle is influenced by the electromagnetic field generated by the other particles and gains or loses energy under the action of the Lorentz force. There is thus a constant exchange of energy between the particles and the field. We shall consider how to assign an 'energy density' to the electromagnetic field in such a way that the total energy of the particles and the field is conserved.

Suppose that in the inertial coordinate system t, x, y, z, there are $\sigma(t, x, y, z)$ per unit volume, and that the particles have velocity $\boldsymbol{u} = \boldsymbol{u}(t, x, y, z)$. Then

$$\rho = e\sigma, \qquad \boldsymbol{J} = e\sigma\boldsymbol{u}\,.$$

By Proposition 8.1, the motion of a particle satisfies

$$\frac{\mathrm{d}}{\mathrm{d}t}\left(m\gamma(u)c^2\right) = e\boldsymbol{E}\,.\,\boldsymbol{u}\,.$$

It follows that between t and $t + \delta t$, the energy $m_I c^2$ of a particle changes by $e\boldsymbol{E}\,.\,\boldsymbol{u}\,\delta t$. If T is the total energy of the particles and V is a fixed volume containing the entire cloud, then

$$\begin{aligned}\frac{\mathrm{d}T}{\mathrm{d}t} &= \int_V e\sigma\boldsymbol{E}\,.\,\boldsymbol{u}\,\mathrm{d}V \\ &= \int_V \boldsymbol{E}\,.\,\boldsymbol{J}\,\mathrm{d}V \\ &= \int_V \frac{1}{\mu_0}\boldsymbol{E}\,.\left(\operatorname{curl}\boldsymbol{B} - \frac{1}{c^2}\frac{\partial \boldsymbol{E}}{\partial t}\right)\mathrm{d}V\end{aligned}$$

by Maxwell's equations. But

$$\boldsymbol{E}\,.\operatorname{curl}\boldsymbol{B} = \operatorname{div}(\boldsymbol{B}\wedge\boldsymbol{E}) + \boldsymbol{B}\,.\operatorname{curl}\boldsymbol{E} = \operatorname{div}(\boldsymbol{B}\wedge\boldsymbol{E}) - \boldsymbol{B}\,.\,\frac{\partial \boldsymbol{B}}{\partial t}\,.$$

Hence

$$\begin{aligned}\frac{\mathrm{d}T}{\mathrm{d}t} &= \int_V \frac{1}{\mu_0}\left[\operatorname{div}(\boldsymbol{B}\wedge\boldsymbol{E}) - \frac{1}{2}\frac{\partial}{\partial t}\left(c^{-2}\boldsymbol{E}\,.\,\boldsymbol{E} + \boldsymbol{B}\,.\,\boldsymbol{B}\right)\right]\mathrm{d}V \\ &= -\int_{\partial V}\frac{1}{\mu_0}(\boldsymbol{E}\wedge\boldsymbol{B})\,.\,\mathrm{d}\boldsymbol{S} - \frac{\mathrm{d}}{\mathrm{d}t}\int_V\left(\frac{\epsilon_0}{2}\boldsymbol{E}\,.\,\boldsymbol{E} + \frac{1}{2\mu_0}\boldsymbol{B}\,.\,\boldsymbol{B}\right)\mathrm{d}V\,.\end{aligned}$$

It makes sense, therefore, to identify the quantity

$$\frac{\epsilon_0}{2}\boldsymbol{E}\,.\,\boldsymbol{E} + \frac{1}{2\mu_0}\boldsymbol{B}\,.\,\boldsymbol{B}$$

with the *energy density* of the electromagnetic field and to identify the vector

$$\frac{\boldsymbol{E}\wedge\boldsymbol{B}}{\mu_0}$$

with the *energy flux* (it is called the *Poynting vector*). Then we can interpret our formula for $\mathrm{d}T/\mathrm{d}t$ as a statement of energy conservation; the increase in the energy of the particles is matched by the flow of energy across the surface of V (the first integral on the right-hand side) and a decrease in the energy of the field within V (the second integral).

Example 8.9

In the case of the linearly polarized monochromatic plane wave

$$\boldsymbol{E} = \boldsymbol{\alpha} \cos \Omega, \qquad c\boldsymbol{B} = \boldsymbol{e} \wedge \boldsymbol{E},$$

where $\Omega = \omega(t - \boldsymbol{e} \,.\, \boldsymbol{r}/c)$, the Poynting vector is

$$\frac{1}{\mu_0} \boldsymbol{E} \wedge \boldsymbol{B} = \frac{\boldsymbol{\alpha} \,.\, \boldsymbol{\alpha} \cos^2 \Omega}{\mu_0 c} \boldsymbol{e} \,.$$

So energy flows in the direction of the vector $\boldsymbol{e}$, as one would expect.

8.6 The Four-momentum of a Photon

The electromagnetic field has an energy density and electromagnetic waves can carry energy between particles. It follows that individual photons carry energy and that four-momentum can be transferred back and forth between particles and electromagnetic fields by absorbtion and emission of photons. In fact, the basic principles of quantum theory imply that a photon of angular frequency ω has energy $E = \hbar\omega$, where

$$\hbar = 1.05 \times 10^{-34}\,\mathrm{Js}$$

is Planck's constant (divided by 2π). If this is to hold in every inertial frame, then a photon with frequency four-vector K must have four-momentum

$$P = \frac{\hbar}{c} K \,, \tag{8.18}$$

since in any frame, the temporal part of P is E/c and the temporal part of K is ω.

The rest mass of an ordinary particle can be found from its four-momentum by using the formula $m^2c^2 = g(P, P)$. In the case of a photon, however, $g(P, P) = 0$ since K is null. Photons are therefore examples of 'zero-rest-mass' particles, although this is a misleading term since photons do not have rest frames—they have speed c in all inertial coordinate systems.

Photon Hypothesis

The law of conservation of four-momentum extends to collisions involving photons.

Problem 8.10 (†Compton Scattering)

A photon of frequency ω collides with an electron of rest mass m, which is initially at rest. After the collision, the photon has frequently ω'. Show that

$$\hbar\omega\omega'(1-\cos\theta) = mc^2(\omega-\omega'), \tag{8.19}$$

where θ is the angle between the initial and final trajectories of the photon.

Solution

In the inertial coordinate system in which the electron is initially at rest, the four-momenta of the electron before and after the collision are

$$P = m(c,0) \quad \text{and} \quad Q = m\gamma(u)(c,\boldsymbol{u}), \tag{8.20}$$

where $\boldsymbol{u}$ is the electron velocity after the collision. The four-momenta of the photon before and after the collision are

$$L = \frac{\hbar\omega}{c}(1,\mathbf{e}) \quad \text{and} \quad L' = \frac{\hbar\omega'}{c}(1,\mathbf{e}'), \tag{8.21}$$

where $\mathbf{e}$ and $\mathbf{e}'$ are unit vectors along the initial and final trajectories of the photon. By conservation of four-momentum $P+L=Q+L'$ and therefore

$$g(L+P,L+P) = g(Q+L',Q+L').$$

But

$$g(P,P) = g(Q,Q) = m^2c^2 \quad \text{and} \quad g(L,L) = g(L',L') = 0.$$

Therefore

$$g(L,P) = g(L',Q).$$

By taking the inner product of the four-momentum conservation equation with L', we also have

$$g(L',Q) = g(L',L) + g(L',P).$$

It follows that

$$g(L',L) = g(L',Q) - g(L',P) = g(P,L-L').$$

But from (8.21),

$$g(L', L) = \frac{\hbar^2 \omega \omega'}{c^2}(1 - e \,.\, e') = \frac{\hbar^2 \omega \omega'}{c^2}(1 - \cos\theta)$$

and

$$g(P, L - L') = m\hbar(\omega - \omega') \,.$$

The result follows. □

8.7 *Advanced and Retarded Solutions

The four-potential form of Maxwell's equations determines the electromagnetic field of a distribution of charges. In particular, they have solutions corresponding to the emission of light or other electromagnetic radiation by moving charges.

Suppose that the charges are contained in some bounded region and that they are at rest relative to an inertial frame before $t = t_0$. Suppose also that before this time, the only electromagnetic field present is that generated by the charges themselves. Then for $t < t_0$, there will be no magnetic field and the electric field will be the sum of the electrostatic fields of the individual charges, which falls off like r^{-2} as $r \to \infty$, where $r = |\boldsymbol{r}|$, $\boldsymbol{r} = (x, y, z)$. So we can take $\boldsymbol{A} = 0$ for $t < t_0$, with ϕ and $\boldsymbol{E}$ falling off like r^{-1} and r^{-2} respectively. Thus

$$\Phi = O(r^{-1}), \quad F = O(r^{-2}), \qquad \text{as } r \to \infty \text{ for } t < t_0. \tag{8.22}$$

We shall show that under this condition, the four-potential at an event A is given by an integral over the past light-cone of A; so the motion of a charge at an event B affects the four-potential at events on the future light-cone of B. Thus Maxwell's equations propagate disturbances along null lines.

Proposition 8.11 (Retarded Potential)

Suppose that Φ satisfies

$$\Box \Phi = \frac{1}{c\epsilon_0} J, \qquad \operatorname{Div} \Phi = 0$$

together with the boundary condition (8.22). Then at the event $(t, \boldsymbol{r}) = 0$

$$\Phi = \frac{1}{4\pi\epsilon_0 c} \int J(-r/c, \boldsymbol{r}) \, \frac{\mathrm{d}V}{r} \,,$$

where the integral is over all space.

Proof

We shall look at only the temporal part, since the proof for the other three components is the same. Define ψ and χ as functions of $\boldsymbol{r}$ by replacing t by $-r/c$ in $\phi(t, \boldsymbol{r})$ and $\partial_t\phi(t, \boldsymbol{r})$, respectively. Then

$$\begin{aligned}
\frac{\partial \psi}{\partial x} &= \frac{\partial \phi}{\partial x} - \frac{x}{cr}\frac{\partial \phi}{\partial t} \\
\frac{\partial^2 \psi}{\partial x^2} &= \frac{\partial^2 \phi}{\partial x^2} - \frac{2x}{cr}\frac{\partial^2 \phi}{\partial x \partial t} + \frac{x^2}{c^2 r^2}\frac{\partial^2 \phi}{\partial t^2} - \frac{1}{cr}\frac{\partial \phi}{\partial t} + \frac{x^2}{cr^3}\frac{\partial \phi}{\partial t} \\
\operatorname{div}\left(\frac{\chi \boldsymbol{r}}{r^2}\right) &= \frac{1}{r^2}\frac{\partial \phi}{\partial t} + \frac{\boldsymbol{r}}{r^2} \cdot \operatorname{grad}\left(\frac{\partial \phi}{\partial t}\right) - \frac{1}{cr}\frac{\partial^2 \phi}{\partial t^2},
\end{aligned}$$

where the right-hand sides are evaluated at $t = -r/c$, with similar expressions for the other derivatives. Therefore

$$\frac{1}{r}\nabla^2\psi + \frac{2}{c}\operatorname{div}\left(\frac{\chi \boldsymbol{r}}{r^2}\right) = -\frac{1}{r}\left(\frac{1}{c^2}\frac{\partial^2 \phi}{\partial t^2} - \nabla^2 \phi\right) = -\frac{\rho(-r/c, \boldsymbol{r})}{r\epsilon_0}.$$

Let S_1 be a small sphere of radius r_1, with its centre at $\boldsymbol{r} = 0$, and let S_2 be a large concentric sphere of radius r_2. By applying the divergence theorem to the volume V between the spheres, and by letting $r_1 \to 0$, we find

$$4\pi\psi(0) = \int_V \frac{\rho(-r/c, \boldsymbol{r})}{r\epsilon_0}\,\mathrm{d}V + \int_{S_2}\left[\frac{2\chi \boldsymbol{r}}{cr^2} + \frac{\operatorname{grad}\psi}{r} - \psi \operatorname{grad}\left(\frac{1}{r}\right)\right] . \mathrm{d}\boldsymbol{S}.$$

As $r_2 \to \infty$, the surface integral goes to zero by (8.22) since $t < t_0$ for large r at the event $(-r/c, \boldsymbol{r})$. The proposition follows. □

Proposition 8.12

The volume element $\mathrm{d}N = \mathrm{d}V/r$ on the past light-cone of the event $(t, \boldsymbol{r}) = 0$ is invariant under Lorentz transformations.

Proof

Since it is clearly invariant under rotations, we have only to show that it is invariant under the standard Lorentz transformation. That is, we must show that if

$$\begin{pmatrix} -r \\ x \\ y \\ z \end{pmatrix} = L_u \begin{pmatrix} -r' \\ x' \\ y' \\ z' \end{pmatrix}$$

where $r' = |\boldsymbol{r}'|$, $\boldsymbol{r}' = (x', y', z')$, and L_u is the standard Lorentz transformation matrix in (5.5), then

$$\begin{vmatrix} \partial_{x'}x & \partial_{y'}x & \partial_{z'}x \\ \partial_{x'}y & \partial_{y'}y & \partial_{z'}y \\ \partial_{x'}z & \partial_{y'}z & \partial_{z'}z \end{vmatrix} = \frac{r}{r'} .$$

But $y = y'$, $z = z'$, and $x = \gamma(u)(x' - ur'/c)$. Therefore

$$\partial_{x'}x = \gamma(u)\left(1 - \frac{ux'}{r'c}\right) = \frac{r}{r'} .$$

The proposition follows. □

Under the boundary condition (8.22), therefore, we can write the four-potential at an event in the form of an integral over the past light-cone N of the event:

$$\Phi = \frac{1}{4\pi\epsilon_0 c}\int_N J\,\mathrm{d}N .$$

This is called the *retarded potential* since the potential is determined by the behaviour of the charges at earlier times. There is an equally valid *advanced potential*, where the integral is over the *future* light cone. Here the boundary condition is the rather more artificial condition that (8.22) holds, but for $t > t_0$. This solution can be understood as the time reversal of the emission of radiation by moving charges; instead of travelling out from the disturbance, the radiation focuses in on the moving charges, and exactly cancels the radiation they would otherwise have emitted, leaving a static field at later times.

EXERCISES

8.1. Show that the *dual electromagnetic field*

$$F^* = \begin{pmatrix} 0 & -cB_1 & -cB_2 & -cB_3 \\ cB_1 & 0 & -E_3 & E_2 \\ cB_2 & E_3 & 0 & -E_1 \\ cB_3 & -E_2 & E_1 & 0 \end{pmatrix}$$

also has the tensor property (i.e. $F'^* = L^{\mathrm{t}}F^*L$).

By considering the trace of gF^*gF, give an alternative proof of the invariance of $\boldsymbol{B}.\boldsymbol{E}$.

8.2. Show that if V is a four-velocity (written as a column vector) and F is an electromagnetic field, then gFV and gF^*V transform as four-vectors. Show that an observer with four-velocity V sees no electric field if $FV = 0$ or no magnetic field if $F^*V = 0$. Show that the former condition is equivalent to

$$\boldsymbol{E}\,.\,\boldsymbol{v} = 0 \qquad \text{and} \qquad \boldsymbol{E} + \boldsymbol{v} \wedge \boldsymbol{B} = 0$$

where $\boldsymbol{v}$ is the observer's three-velocity. Deduce that there is a frame in which the electric field vanishes if and only if

$$\boldsymbol{E}\,.\,\boldsymbol{B} = 0 \qquad \text{and} \qquad \boldsymbol{E}\,.\,\boldsymbol{E} < c^2\boldsymbol{B}\,.\,\boldsymbol{B}\,.$$

Under what conditions does there exist a frame in which the magnetic field vanishes?

8.3. Let P and K be constant four-vectors and let X denote the displacement four-vector from some fixed origin. Put $\Omega = g(K, X)/c$. Show that

$$\Phi = P\,\cos(\Omega)$$

is a four-potential for an electromagnetic field without sources in the Lorenz gauge if and only if K is null and orthogonal to P. Find and interpret the corresponding electric and magnetic fields.

8.4. In an inertial frame, the electric field is uniform and constant, and the magnetic field vanishes. A particle of charge e and rest mass m has initial velocity $\boldsymbol{v}_0$ perpendicular to $\boldsymbol{E}$. Show that

$$\boldsymbol{E}\,.\,\boldsymbol{v} = c\boldsymbol{E}\tanh\left(\frac{eE\tau}{mc}\right) \qquad \text{and} \qquad \gamma(v) = \gamma(v_0)\cosh\left(\frac{eE\tau}{mc}\right),$$

where $\boldsymbol{v}$ is the velocity of the particle relative to the frame, τ is proper time, and $E = |\boldsymbol{E}|$. Hence find the trajectory of the particle. In the classical theory, the speed of the particle increases without limit. What happens here?

8.5. A particle of rest mass m, charge e, and initial velocity $\boldsymbol{v}$ relative to some inertial coordinate system is moving in a constant magnetic field $\boldsymbol{B}$ perpendicular to $\boldsymbol{v}$. Show that the particle moves in a circle and that the proper time on the particle worldline that elapses on each circuit is $2\pi m/eB$.

Deduce that if a particle of rest mass m and charge e moves through a constant uniform electromagnetic field in which

$$\boldsymbol{B}.\boldsymbol{E} = 0 \qquad \text{and} \qquad \boldsymbol{E}.\boldsymbol{E} < c^2\boldsymbol{B}.\boldsymbol{B}\,,$$

then after the elapse of proper time $2\pi mc/e\sqrt{c^2\boldsymbol{B}.\boldsymbol{B} - \boldsymbol{E}.\boldsymbol{E}}$ measured along the worldline of the particle, its velocity relative to the inertial coordinate system is equal to its initial velocity.

8.6. Two particles of equal rest mass collide and annihilate each other, producing two photons. Initially one particle is at rest and the other has velocity $\boldsymbol{u}$. Show that if the angles between $\boldsymbol{u}$ and the photon trajectories are θ and ϕ, then

$$\frac{1+\cos(\theta+\phi)}{\cos\theta+\cos\phi}=\sqrt{\frac{\gamma(u)-1}{\gamma(u)+1}}\,.$$

8.7. A particle accelerates by absorbing photons from a parallel beam of light. Relative to an inertial frame, the particle's velocity is $\boldsymbol{v}$. Show that

$$m\gamma(v)(c,\boldsymbol{v})=m_0\gamma(v_0)(v\boldsymbol{v}_0)+\lambda(1,\boldsymbol{k})$$

where m is the particle's rest mass, $\boldsymbol{k}$ is a unit three-vector in the direction of the beam, λ is a function of m, and m_0 and $\boldsymbol{v}_0$ are the initial values of m and $\boldsymbol{v}$. Hence show that

$$c-\boldsymbol{v}\,.\,\boldsymbol{k}=\frac{A}{m^2+B}\qquad\text{and}\qquad c^2-v^2=\frac{2Acm^2}{(m^2+B)^2}\,,$$

where A and B are constants that should be determined from the initial conditions. Show that $m\to\infty$ as $v\to c$.

8.8. $^{+}$A photon of frequency ω travels in the direction of a unit vector $\boldsymbol{e}$ and collides with a stationary electron of rest mass m. Afterwards, the electron travels in a direction making an angle $\theta>0$ with $\boldsymbol{e}$. Show that if $k=\hbar\omega/mc^2$, then

$$\tan\phi=\frac{\sin\theta}{(1+k)(1-\cos\theta)}\,.$$

9
*Tensors and Isometries

Up to this point, we have developed Einstein's special theory of relativity by exposing the inconsistency between electrodynamics and Galilean relativity, and then by establishing the Lorentz invariance of Maxwell's equations. The mathematical structure has emerged piecemeal as we have explored the consequences of operational definitions and isolated basic facts about the observations made by inertial observers. From one point of view, it would have been more straightforward simply to have laid out the structure of Minkowski space, and then to have derived the physical predictions of the theory. The result would have been a more coherent mathematical exposition, but a correspondingly obscure account of the arguments for accepting special relativity as a description of the real world; however impressive the correspondence between prediction and observation, it is difficult without following the argument from first principles to accept a radical departure from intuitive notions of space and time that appear to be soundly based on everyday experience.

This final chapter has a different style from the others. It describes some of the structures that emerge from relativity from a more overtly geometric perspective, so as to allow connections to be made with other parts of mathematics. It explores the connection between relativistic transformation rules—the tensor property—and the symmetries of Minkowski space.

9.1 Affine Space

The starting point will be the description of Minkowski space $\mathbb{M}$ as an *affine space*, modelled on the space of four-vectors, which is a four-dimensional real vector space $\mathbb{V}$ with an indefinite inner product g of signature $+, -, -, -$. That is, $\mathbb{M}$ is a set together with a map $\rho : \mathbb{M} \times \mathbb{V} \to \mathbb{M}$ with the properties

- for all $A \in \mathbb{M}$ and for all $X, Y \in \mathbb{V}$,

$$\rho\big(\rho(A, X), Y\big) = \rho(A, X + Y)\,;$$

- for all $A \in \mathbb{M}$,

$$\rho(A, 0) = A\,;$$

- for all $A, B \in \mathbb{M}$, there exists a unique $X \in \mathbb{V}$ such that

$$\rho(A, X) = B\,.$$

If $\rho(A, X) = B$, then X is called the *displacement four-vector* from A to B. Given a choice of *origin* $A \in \mathbb{M}$, each event $B \in \mathbb{M}$ can be uniquely labelled by the displacement four-vector X from A to B. So we can set up a coordinate system x^0, x^1, x^2, x^3 on $\mathbb{M}$ by choosing an origin $A \in \mathbb{M}$ and an orthonormal basis in $\mathbb{V}$. That is, a basis in which

$$g(V, V) = (V^0)^2 - (V^1)^2 - (V^1)^2 - (V^1)^2\,,$$

where the V^as are the components of $V \in \mathbb{V}$. The coordinates of $B \in \mathbb{M}$ are then defined to be the components x^a in the basis of the displacement four-vector X from A to B. Two such coordinate systems are related by an affine transformation,

$$x^a = \sum_{b=0}^{3} L^a{}_b x'^b + T^a$$

in which $L = \left(L^a{}_b\right)$ is pseudo-orthogonal and $T = (T^a)$ is the displacement four-vector from one origin to the other.

9.2 The Lorentz Group

The *Lorentz group* $\mathbb{L}$ is the group linear transformations $\lambda : \mathbb{V} \to \mathbb{V}$ with the property

$$g(\lambda X, \lambda Y) = g(X, Y) \tag{9.1}$$

for every $X, Y \in \mathbb{V}$. If we choose a pseudo-orthonormal basis, then λ will be given by its linear action on four-vector components:

$$\begin{pmatrix} X^0 \\ X^1 \\ X^2 \\ X^3 \end{pmatrix} \mapsto L \begin{pmatrix} X^0 \\ X^1 \\ X^2 \\ X^3 \end{pmatrix},$$

where L is the matrix of λ. The condition (9.1) is the condition that L should be pseudo-orthogonal.

We have seen the Lorentz group already as a group of coordinate transformations, but we are now looking at it in a different way, as a group of linear transformations of $\mathbb{V}$. That is, we have shifted from the *passive* picture, in which Lorentz transformations change the coordinates labelling events, to the *active* one, in which they act on four-vectors. If we change the pseudo-orthonormal basis in $\mathbb{V}$, then the same transformation λ is represented by the conjugate matrix $M^{-1}LM$, where M is the matrix that determines the change of basis.

Given $\lambda \in \mathbb{L}$, we put

$$\sigma(\lambda) = \operatorname{sign}\left(\frac{g(T, \lambda T')}{g(T, T')}\right),$$

where $T, T' \in \mathbb{V}$ are timelike, and sign takes the values ± 1 as its argument is positive or negative. This is well-defined since T, T' and $\lambda T'$ are timelike, and since the inner product of two timelike four-vectors is necessarily nonzero. It is independent of T and T' because it is unchanged when T is replaced by $-T$ or T' by $-T'$, and because it varies continuously with T and T'. Moreover,

$$\sigma(\lambda\lambda') = \sigma(\lambda)\sigma(\lambda'), \qquad \lambda, \lambda' \in \mathbb{L}.$$

Thus σ is a homomorphism $\mathbb{L} \to \mathbb{Z}_2$. The determinant $\lambda \mapsto \det(\lambda)$ is another such homomorphism.

As a topological space—a subspace of the set of 4×4 matrices—the Lorentz group has four connected components, characterized by the values of σ and det. The *proper orthochronous Lorentz group* $\mathbb{L}_+$ is the subgroup on which $\sigma(\lambda) = \det(\lambda) = 1$; equivalently, it is the subgroup on which the corresponding Lorentz matrices are proper and orthochronous. The Lorentz group differs from the orthogonal group in four, and other, dimensions in that the orthogonal group has only two components, corresponding to the two choices for the sign of the determinant.

Exercise 9.1

Show that $\sigma(\lambda)$ is independent of the choice of T and T'.

The space–time of special relativity is Minkowski space, together with a choice for the 'arrow of time' and for the 'handedness' of space; that is, a choice of a family of pseudo-orthonormal bases in $\mathbb{V}$ related by proper orthochronous Lorentz transformations. The corresponding coordinate systems on $\mathbb{M}$ are the *inertial coordinate systems*.

9.3 Tensors

Four-vectors and electromagnetic fields are examples of *tensors*. More generally, tensors are objects with components that transform in a particular linear way under change of inertial coordinates, as do the components of a four-vector or of an electromagnetic field. From another point of view, tensors are elements of certain *representation spaces* of the Lorentz group.

Definition 9.1

Let U be a real vector space and let $\mathrm{GL}(U)$ denote the group of linear transformations of U. A (linear) *representation* of the Lorentz group on U is a homomorphism $\mathbb{L} \to \mathrm{GL}(U)$.

Speaking more loosely, we call U a *representation space*, with the homomorphism understood. The tensor spaces are natural representation spaces of the Lorentz group. Their construction may seem uncomfortably abstract when compared with the concrete statement of component transformation properties. But the abstract viewpoint reflects the operational definitions of the corresponding physical quantities. An electromagnetic field is defined operationally in terms of the acceleration of a charge; this is what is measured directly, not the individual components of $\boldsymbol{B}$ and $\boldsymbol{E}$ (the entries in F). From this operational point of view, therefore, it is natural to think of the electromagnetic field at an event as a linear map that determines the force on a charged particle from its velocity, by the Lorentz force law. So a natural answer to the question 'what kind of mathematical object is an electromagnetic field?' is that it is a tensor in the sense of the algebraic definition below, which is phrased in terms of linear maps.

To define the tensor spaces, we shall work first in the general context of finite dimensional linear algebra before specializing to the space of four-vectors. We shall use a definition of the tensor product that is convenient for our particular purpose, but which is not entirely standard.[1]

[1] It is equivalent to the standard one; see Proposition 2.7 of [4].

9.4 The Tensor Product

First we recall the definition of the dual space. Let V be a finite-dimensional real vector space, of dimension n. Then the *dual space* V^* is the vector space of linear maps or *covectors*

$$\alpha : V \to \mathbb{R} : v \mapsto \alpha(v) .$$

It also has dimension n. If we choose a basis $\{e_a\} \subset V$ $(a = 1, \ldots, n)$, then the *components* of $\alpha \in V^*$ are the real numbers

$$\alpha_a = \alpha(e_a) .$$

The components $(\alpha_0, \ldots, \alpha_n)$ uniquely determine α, and so we can define a basis V^*, the *dual basis* to the basis in V, by picking out the covectors with components

$$(1, 0, 0, \ldots, 0), \quad (0, 1, 0, \ldots 0), \quad \ldots, \quad (0, 0, 0, \ldots, 1) .$$

A fundamental fact is that there is a natural isomorphism $(V^*)^* \to V$, which is used to identify the two spaces. Under the identification, $v \in V$ becomes the linear map

$$v : V^* \to \mathbb{R} : \alpha \mapsto \alpha(v) .$$

Thus we have a symmetric picture in which elements of V^* are linear maps $V \to \mathbb{R}$, and elements of V are linear maps $V^* \to \mathbb{R}$. The tensor spaces are built up by looking more widely at linear maps from V or V^* to other vector spaces. Their elements are called *tensors*.

Definition 9.2

Let U and V be finite-dimensional real vector spaces. The *tensor product* $U \otimes V$ is the vector space of linear maps $T : V^* \to U$.

In particular, $\mathbb{R} \otimes V^* = V^*$, so we have extended the definition of the dual space. If $u \in U$, $v \in V$, then we denote by $u \otimes v \in U \otimes V$ the linear map

$$u \otimes v : V^* \to U : \beta \mapsto \beta(v)u ,$$

for $\beta \in V^*$.

EXERCISES

9.2. Show that $\mathbb{R}^2 \otimes V = V \oplus V$.

9.3. Show that for any $u_1, u_2 \in U$, $v_1, v_2 \in V$, and $a, b \in \mathbb{R}$,

$$\begin{aligned}(au_1 + bu_2) \otimes v &= au_1 \otimes v + bu_2 \otimes v \\ u \otimes (av_1 + bv_2) &= au \otimes v_1 + bu \otimes v_2 .\end{aligned}$$

Let us choose bases $\{u_i\} \subset U$ and $\{v_a\} \subset V$, where i runs from 1 to $m = \dim U$ and a runs from 1 to $n = \dim V$. Then the components of a covector $\beta \in V^*$ in the dual basis in V^* are

$$\beta_a = \beta(v_a)$$

and a tensor $T \in U \otimes V$ can be represented uniquely by its matrix as a linear map $V^* \to U$. If we denote the entries in the matrix by T^{ia}, then

$$T(\beta) = \sum_{i=1}^{m} \sum_{a=1}^{n} T^{ia} \beta_a u_i ,$$

or, alternatively,

$$T = \sum_{i=1}^{m} \sum_{a=1}^{n} T^{ia} u_i \otimes v_a .$$

Thus $\{u_i \otimes v_a\}$ is a basis for $U \otimes V$ and the entries in the matrix of T are its *components* in this basis. Moreover,

$$\dim (U \otimes V) = mn = \dim U \dim V .$$

The basic properties of the tensor product are given by the following.

Proposition 9.3

For any finite-dimensional vector space U, V, W, there are natural isomorphisms

(i) $U \otimes V \to V \otimes U$,

(ii) $U^* \otimes V^* \to (U \otimes V)^*$,

(iii) $U \otimes (V \otimes W) \to (U \otimes V) \otimes W$.

Proof

The first isomorphism is familiar from elementary linear algebra. It maps $T : V^* \to U$ to its *adjoint* or *dual*, which is the linear map $T^* : U^* \to V$ characterized by

$$\alpha(T(\beta)) = \beta(T^*(\alpha))$$

for every $\alpha \in U^*$, $\beta \in V^*$. So the isomorphism sends

$$u \otimes v \in U \otimes V \quad \text{to} \quad v \otimes u \in V \otimes U\,.$$

If $T \in U \otimes V$ has components T^{ia}, then the corresponding tensor in $V \otimes U$ has components T^{ai}.

The second isomorphism maps $\alpha \otimes \beta$ to $\gamma \in (U \otimes V)^*$, where

$$\gamma(u \otimes v) = \alpha(u)\,\beta(v)\,.$$

Elements of the form $u \otimes v$ span $U \otimes V$, so this, together with linearity, determines γ uniquely; and since the elements of the form $\alpha \otimes \beta$ span $U^* \otimes V^*$, the requirement that $\alpha \otimes \beta$ should map to γ determines the isomorphism uniquely. But one must check at both steps that the definitions are good. That is, if $T \in U \otimes V$ is decomposed in two different ways as a linear combination of tensors of the form $u \otimes v$, then we get the same answer for $\gamma(T)$ from either decomposition. Similarly that the image in $(U \otimes V)^*$ of $\omega \in U^* \otimes V^*$ does not depend on the way in which is ω written as a linear combination of elements of the form $\alpha \otimes \beta$. These tasks are left as an exercise.

The third isomorphism is determined in an analogous way by the condition

$$u \otimes (v \otimes w) \mapsto (u \otimes v) \otimes w$$

for $u \in U$, $v \in V$, $w \in W$. □

The third isomorphism in the proposition allows us to write the triple tensor product of three spaces as $U \otimes V \otimes W$, without specifying the order in which the products are taken. Some care is necessary, however. It *is* true that if $u \in U$, $v \in V$, $w \in W$, then

$$(u \otimes v) \otimes w = u \otimes (v \otimes w)$$

in $U \otimes V \otimes W$, so it is legitimate to omit the brackets and denote both sides by $u \otimes v \otimes w$. Also, if $U = V = W$, then

$$(u + v) \otimes w = u \otimes w + v \otimes w$$

in $U \otimes U$. But in general

$$u \otimes v \neq v \otimes u\,.$$

9.5 Tensors in Minkowski Space

Tensors in Minkowski space are elements of the various tensor products of $\mathbb{V}$, the space of four-vectors, with itself and with its dual space $\mathbb{V}^*$, the space of covectors.

Definition 9.4

A *tensor of type* (p, q), for non-negative integers p and q, is an element of the vector space $\mathbb{V}^p_q$, defined by

$$\mathbb{V}^p_q = \overbrace{\mathbb{V} \otimes \cdots \otimes \mathbb{V}}^{p} \otimes \underbrace{\mathbb{V}^* \otimes \cdots \otimes \mathbb{V}^*}_{q} .$$

A *tensor field* is a map $T : \mathbb{M} \to \mathbb{V}^p_q$.

When $p = q = 0$, the tensors of type (p, q) are real numbers—that is, scalars—and the corresponding tensor fields are simply real-valued functions on space–time. Two other special cases are $\mathbb{V}^1_0 = V$ and $\mathbb{V}^0_1 = V^*$. Tensors and tensor fields can be added to tensors or tensor fields of the same type; and if $T \in \mathbb{V}^p_q$ and $T' \in \mathbb{V}^{p'}_{q'}$, then

$$T \otimes T' \in \mathbb{V}^{p+p'}_{q+q'} .$$

Thus far, the theory of tensors makes sense in any affine space. In Minkowski space there is an additional set of operations defined in terms of the (indefinite) inner product. This determines an isomorphism $g : \mathbb{V} \to \mathbb{V}^*$ by mapping $X \in \mathbb{V}$ to the linear map

$$\mathbb{V} \to \mathbb{R} : Y \mapsto g(X, Y) . \tag{9.2}$$

Thus g is a tensor of type (0,2), which is called the *metric tensor*. The inverse g^{-1} is an isomorphism $\mathbb{V}^* \to \mathbb{V}$, and therefore a tensor of type (2,0).

By applying g to one or other of the first p factors in

$$\mathbb{V}^p_q = \mathbb{V} \otimes \cdots \otimes \mathbb{V} \otimes \mathbb{V}^* \otimes \cdots \otimes \mathbb{V}^* ,$$

we get a sequence of isomorphisms $\mathbb{V}^p_q \to \mathbb{V}^{p-1}_{q+1}$. If it is applied to the rth factor, then the isomorphism is called 'lowering the rth index', for reasons that will be apparent when we look at the effect on components. We can similarly apply g^{-1} to one of the $\mathbb{V}^*$ factors to get an 'index raising' operation, that is an isomorphism $\mathbb{V}^p_q \to \mathbb{V}^{p+1}_{q-1}$. In early treatments of special relativity, it was conventional to identify different types of tensor, and so not to make a distinction between the spaces $\mathbb{V}^p_q$ for the same value of $p + q$. But this is unhelpful when the theory of tensors is extended in the context of differential geometry or general relativity.

9.6 Tensor Components

The choice of inertial coordinate system determines a basis in $\mathbb{V}$, made up of the vectors with components

$$(1,0,0,0), \quad (0,1,0,0), \quad (0,0,1,0), \quad \text{and} \quad (0,0,0,1).$$

We shall denote these by v_0, v_1, v_2, v_3, where the subscripts label the different vectors in the basis, not their individual components. Any other four-vector X with components X^a can be expanded in the basis by writing

$$X = \sum_0^3 X^a v_a .$$

The coordinate system also determines a dual basis $\nu^0, \nu^1, \nu^2, \nu^3$ in $\mathbb{V}^*$, with the defining property that

$$\nu^a(v_b) = \delta^a_b ,$$

where δ^a_b is the Kronecker delta, equal to one when $a = b$, and to zero otherwise. Again the indices label the different components. The use of subscripts to label the elements of the basis and of superscripts to label the elements of the dual basis is designed to allow Einstein's index conventions to work in a consistent way, as we see shortly.

An element $\alpha \in \mathbb{V}^*$ can similarly be expanded in terms of its components as

$$\alpha = \sum_0^3 \alpha_a \nu^a .$$

In order to avoid an explosion of summation signs, which can obstruct a clear understanding of tensor equations, we shall follow Einstein in adopting the *summation convention*; whenever an index is repeated, once as an upper index and once as a lower index, a sum is implied over the range $0, 1, 2, 3$; and any equation involving tensor components is required to hold for all values of its *free indices*—indices that are not repeated and over which there is no sum—in the same range. Thus we can write

$$X = X^a v_a \qquad \text{and} \qquad \alpha = \alpha_a \nu^a$$

and also

$$X^a = \nu^a(X) \qquad \text{and} \qquad \alpha_a = \alpha(v_a) .$$

The inner product takes the form

$$g(X, Y) = g_{ab} X^a Y^b ,$$

where

$$g_{00} = -g_{11} = -g_{22} = -g_{33} = 1$$

and $g_{ab} = 0$ for $a \neq b$.

The convention has its pitfalls; one must always keep the summations in mind, otherwise there is a temptation to cancel terms that appear to be common factors on the two sides of an equation, but in fact are not. But it also has inbuilt checks; both the letters and positions of the free indices on the two sides of an equation must match; and the summations are always over a single upper index and a corresponding single lower index.

A tensor T of type p, q can be expanded in the form

$$T = T^{ab\ldots c}{}_{de\ldots f}\, v_a \otimes v_b \otimes \cdots \otimes v_c \otimes \nu^d \otimes \nu^e \otimes \cdots \otimes \nu^f ,$$

with, of course, summation over all the indices on the right. The numbers

$$T^{ab\ldots c}{}_{de\ldots f}$$

are the *components* of T in the coordinate system.

When we make the transformation (5.10) from one inertial coordinate system to another, we have to change the bases to new ones v'_a and ν'^a. From the transformation rules for four-vector and covector components, we can see that these are related to the originals by

$$v'_b = v_a L^a{}_b \qquad \text{and} \qquad \nu^a = L^a_b \nu'^b . \tag{9.3}$$

Note carefully on which sides the original and new basis elements appear here in the two equations. They are consistent since from the first equation

$$\nu^c(v'_b) = \nu^c(v_a) L^a{}_b = \delta^c_a L^a{}_b = L^c{}_b$$

while from the second we also have

$$\nu^c(v'_b) = L^c{}_d \nu'^d(v'_b) = L^c{}_d \delta^d_b = L^c{}_b .$$

It follows from (9.3) that the components of a tensor in the two coordinate systems are related by the transformation rule

$$T^{a\ldots c}{}_{e\ldots f} L^e{}_p \ldots L^f{}_r = L^a{}_s \ldots L^c{}_u T'^{s\ldots u}{}_{p\ldots r} .$$

This determines the new components in terms of the old, because the Lorentz matrices are invertible and so can be transferred from one side to the other. It generalizes the transformation rules for four-vector and covector components; and it can also be used to *characterize* a tensor: it justifies the older definition of a tensor as an array of numbers (the components) with a certain transformation law under change of coordinates.

9.7 Examples of Tensors

The Kronecker Delta

The identity map $\mathbb{V} \to \mathbb{V}$ is a tensor of type $(1,1)$. Its components are δ^a_b, where δ^a_b is equal to one when $a = b$ and to zero otherwise.

The Metric

The metric tensor has components g_{ab} in any inertial coordinate system. The inverse is a tensor of type $(2,0)$. Its components have the same numerical values as those of the metric itself, since the inverse of a diagonal matrix with ones and minus ones on the diagonal is the same matrix, but they are denoted now by g^{ab} so that the index conventions can be applied consistently.

Contraction, Raising, and Lowering

If $T : \mathbb{V} \to \mathbb{V}$ is a tensor of type $(1,1)$, then its trace $\mathrm{tr}(T)$ is a scalar (a real number), given by

$$\mathrm{tr}(T) = T^a{}_a \, .$$

This extends in a natural way to a linear map that sends a tensor T is of type (p,q) to its *contraction*, which is a tensor of type $(p-1, q-1)$, with components

$$T^{ab\ldots c}{}_{ad\ldots e} \, .$$

By combining contraction with tensor multiplication by g or g^{-1}, we obtain the *lowering* and *raising* operators that map tensors of type (p,q) to tensors of the respective types $(p-1, q+1)$ and $(p+1, q-1)$. Their components are denoted by

$$T_a{}^{b\ldots c}{}_{pq\ldots s} = g_{ai} T^{ib\ldots c}{}_{pq\ldots s}$$

and

$$T^{a\ldots cd}{}_{p\ldots s} = g^{di} T^{ab\ldots c}{}_{ip\ldots s} \, .$$

These are in fact the operations of lowering the first upper index and raising the first lower index; by combining the operations with permutations, one can raise and lower other indices. Because of the existence of these operations, it is conventional to use staggering to keep track of the order amongst all indices, upper and lower.

Gradient Four-vector

The gradient covector of a function f on space–time is the covector field ∂f defined at $E \in \mathbb{M}$ by

$$\partial f(X) = \frac{\mathrm{d}}{\mathrm{d}s}\Big(f\big(\rho(E, sX)\big)\Big)\Big|_{s=0}$$

for any four-vector X. That is, $\partial f(X)$ is the derivative of f along X. By the chain rule,

$$\partial f(X) = X^a \partial_a f\,,$$

where $\partial_a = \partial/\partial x^a$, so ∂f has components $\partial_a f$.

By raising the index, we obtain the *gradient four-vector* $\operatorname{Grad} f$. It has components $g^{ab}\partial_b f$; that is,

$$\left(\frac{1}{c}\frac{\partial f}{\partial t}, -\frac{\partial f}{\partial x}, -\frac{\partial f}{\partial y}, -\frac{\partial f}{\partial z}\right).$$

The gradient extends to tensors, so that if T is of type (p, q), then we can define a tensor field ∇T of type $(p, q+1)$, called the *covariant derivative* of T, with components

$$\nabla_a T^{b\ldots d}{}_{e\ldots f} = \partial_a T^{b\ldots d}{}_{e\ldots f}\,.$$

Invariant Operators

The invariant operators defined in §5.12 are given by combining ∇ with raising, lowering and contraction. Thus

$$\operatorname{Div} X = \nabla_a X^a, \qquad \Box = \nabla_a \nabla^a\,,$$

and $\nabla_X Y$ has components $X^a \nabla_a Y^b$. Invariance follows from the tensorial character of the operations.

The Electromagnetic Field Tensor

The electromagnetic field F is a tensor of type $(0, 2)$, with components F_{ab} given by

$$(F_{ab}) = \begin{pmatrix} 0 & E_1 & E_2 & E_3 \\ -E_1 & 0 & -cB_3 & cB_2 \\ -E_2 & cB_3 & 0 & -cB_1 \\ -E_3 & -cB_2 & cB_1 & 0 \end{pmatrix}.$$

As an abstract tensor, F is a linear map $\mathbb{V} \to \mathbb{V}^*$. If V is the four-velocity of a particle with charge e and rest mass m, then the covector $f = eF(V)/c$, with components

$$f_a = \frac{e}{c} F_{ab} V^b ,$$

is called the *Lorentz four-force*. It determines the particle's motion by the energy equation

$$\frac{\mathrm{d}E}{\mathrm{d}\tau} = f(U) \tag{9.4}$$

where U is the four-velocity of an observer, $E = g(U, P)$ is the energy relative to the observer, $P = mV$ is the four-momentum, and τ is the proper time along the particle's worldline. That Equation (9.4) holds for every choice of observer is equivalent to the equation of motion

$$m\frac{\mathrm{d}^2 x^a}{\mathrm{d}\tau^2} = \frac{e}{c} F^a_{\ b} \frac{\mathrm{d}x^b}{\mathrm{d}\tau} ,$$

which in turn is a restatement of Proposition 8.1 in tensor notation.

Maxwell's Equations

By lowering the index on the four-potential Φ we obtain a covector field with components

$$(\Phi_a) = (\phi,\ -cA_1,\ -cA_2,\ -cA_3) ,$$

in terms of which the electromagnetic field is given by

$$F_{ab} = \nabla_a \Phi_b - \nabla_b \Phi_a . \tag{9.5}$$

For example, since $\nabla_a = \partial_a$,

$$E_1 = F_{01} = \partial_0 \Phi_1 - \partial_1 \Phi_0 = -\frac{\partial A_1}{\partial t} - \frac{\partial \phi}{\partial x}$$

and

$$B_1 = \frac{F_{32}}{c} = -\partial_3 A_2 + \partial_2 A_3 .$$

It follows from (9.5) that

$$\nabla_a F_{bc} = \nabla_a \nabla_b \Phi_c - \nabla_a \nabla_c \Phi_b ,$$

and hence that

$$\nabla_a F_{bc} + \nabla_b F_{ca} + \nabla_c F_{ab} = 0 ,$$

since $\nabla_a \nabla_b = \nabla_b \nabla_a = \partial_a \partial_b - \partial_b \partial_a = 0$. By choosing Φ to be in the Lorenz gauge, and by contracting over a, b, we have

$$\nabla_a F^{ab} = \Box \Phi^b - \nabla^b \nabla_a \Phi^a = \frac{1}{\epsilon_0 c} J^b ,$$

where $F^{ab} = g^{ac} g^{bd} F_{cd}$ and we have used the gauge condition $\nabla_a \Phi^a = 0$.

EXERCISES

9.4. Show that Maxwell's equations are equivalent to

$$\nabla_a F_{bc} + \nabla_b F_{ca} + \nabla_c F_{ab} = 0, \qquad \nabla_a F^{ab} = \frac{1}{\epsilon_0 c} J^b .$$

Show that F_{ab} is unchanged when Φ_a is replaced by $\Phi_a + \partial_a u$ for some function u of the space–time coordinates.

9.5. Show that

$$T_{ab} = \epsilon_0 \left(F_{ad} F^d_{\ b} + \tfrac{1}{4} g_{ab} F_{de} F^{de} \right) \tag{9.6}$$

are the components of a tensor T (it is called the *electromagnetic energy–momentum tensor*). Show that in any inertial coordinate system, T_{00} is the energy density of the electromagnetic field. Show that in the absence of sources,

$$\nabla_a T^{ab} = 0 .$$

Write this equation in three-vector notation, and interpret it in terms of energy conservation.

9.6. Show that if N is an eigenvector of F with nonzero eigenvalue s, that is $F_{ab} N^b = s N_a$, then N is null. By writing $N = (1, \boldsymbol{e})$, show that the eigenvalue equation is equivalent to

$$\boldsymbol{E} . \boldsymbol{e} = s, \qquad \boldsymbol{E} + c\boldsymbol{e} \wedge \boldsymbol{B} = s\boldsymbol{e} .$$

Show that in general F has two independent real null eigenvectors. What are the special cases? Show that if an observer's four-velocity is a sum of two null eigenvectors of F, then the electric and magnetic fields seen by the observer will be proportional.

Duality

The determinant $\det F$ of the matrix of components of an electromagnetic field is invariant under Lorentz transformations, and is therefore invariant. It is a homogeneous quartic in the entries in F, but since F is skew-symmetric, it turns out to be the square of a homogeneous quadratic $\mathrm{Pf}\, F$, which is called the *Pfaffian* of F. In fact,

$$\det F = c^2 (\boldsymbol{B} . \boldsymbol{E})^2$$

and so

$$\mathrm{Pf}\, F = c\boldsymbol{B} . \boldsymbol{E}$$

up to sign—the sign can be chosen consistently provided that one allows only proper orthochronous Lorentz transformations. The Pfaffian is an invariant quadratic form on the space of skew-symmetric tensors of type $(0,2)$, and therefore determines a natural linear map to $F \mapsto F^*$ to the dual space of tensors of skew-symmetric type $(2,0)$. By lowering the indices, one obtains a new skew tensor of type $(0,2)$, which is the dual electromagnetic field that appeared in Exercise 8.1. So the invariance of the Pfaffian implies that the construction of the dual field is independent of the choice of inertial coordinate system, and hence that the dual has the tensor property.

Exercise 9.7

Show that the tensor in (9.6) can be written in the form

$$T_a{}^b = \frac{\epsilon_0}{2}\left(F_{ad}F^{db} + F^*_{ad}F^{*db}\right).$$

9.8 One-parameter Subgroups

Much can be learned about the Lorentz group by studying its one-parameter subgroups.

Definition 9.5

A *one-parameter subgroup* of a group G is a homomorphism $\mathbb{R} \to G : t \mapsto \lambda_s$, where $\mathbb{R}$ is regarded as a group under addition.

The *generator* of a one-parameter subgroup of the Lorentz group is the linear map $\gamma : \mathbb{V} \to \mathbb{V}$ defined by

$$\gamma X = \frac{\mathrm{d}}{\mathrm{d}s}(\lambda_s X)\bigg|_{s=0}.$$

If λ_s has matrix representation L_s in some inertial coordinate system, then γ has matrix representation G, where G is the derivative of L_s at $s = 0$. By differentiating the Lorentz condition

$$g(\lambda_s X, \lambda_s Y) = g(X, Y)$$

at $s = 0$, we obtain

$$g(\gamma X, Y) + g(X, \gamma Y) = 0 \tag{9.7}$$

for every $X, Y \in \mathbb{V}$. Therefore

$$g_{ac}G^c{}_b + g_{cb}G^c{}_a = 0$$

where $G = \left(G^a{}_b\right)$. Consequently, gG is a skew-symmetric matrix, and so G must be of the form

$$G = \left(G^a{}_b\right) = \begin{pmatrix} 0 & e_1 & e_2 & e_3 \\ e_1 & 0 & b_3 & -b_2 \\ e_2 & -b_3 & 0 & b_1 \\ e_3 & b_2 & -b_1 & 0 \end{pmatrix}. \tag{9.8}$$

When the inertial coordinate system is changed, G is replaced by its conjugate $M^{-1}GM$, where M is the Lorentz matrix appearing in the coordinate transformation.

Exercise 9.8

Show that this form for G is preserved when the inertial coordinate system is changed.

Two one-parameter subgroups of the Lorentz group have appeared so far. The first is the group of rotations in the x, y plane, where

$$L_\theta = \begin{pmatrix} 1 & 0 & 0 & 0 \\ 0 & \cos\theta & \sin\theta & 0 \\ 0 & -\sin\theta & \cos\theta & 0 \\ 0 & 0 & 0 & 1 \end{pmatrix}, \qquad G = \begin{pmatrix} 0 & 0 & 0 & 0 \\ 0 & 0 & 1 & 0 \\ 0 & -1 & 0 & 0 \\ 0 & 0 & 0 & 0 \end{pmatrix}.$$

The second is the group of standard Lorentz transformations, where

$$L_\phi = \begin{pmatrix} \cosh\phi & \sinh\phi & 0 & 0 \\ \sinh\phi & \cosh\phi & 0 & 0 \\ 0 & 0 & 1 & 0 \\ 0 & 0 & 0 & 1 \end{pmatrix}, \qquad G = \begin{pmatrix} 0 & 1 & 0 & 0 \\ 1 & 0 & 0 & 0 \\ 0 & 0 & 0 & 0 \\ 0 & 0 & 0 & 0 \end{pmatrix}.$$

In both cases,

$$L_s L_{s'} = L_{s+s'} .$$

In the first case, the parameter is the angle of rotation $s = \theta$. In the second, $s = \phi$ is the rapidity. More generally, we can combine a Lorentz transformation in the t, z coordinates with a rotation in the x, y plane to obtain the one-parameter subgroup

$$L_s = \begin{pmatrix} \cosh as & 0 & 0 & \sinh as \\ 0 & \cos bs & \sin bs & 0 \\ 0 & -\sin bs & \cos bs & 0 \\ \sinh as & 0 & 0 & \cosh as \end{pmatrix} \tag{9.9}$$

where s is the parameter and a and b are constant. Here the generator is

$$G = \begin{pmatrix} 0 & 0 & 0 & a \\ 0 & 0 & b & 0 \\ 0 & -b & 0 & 0 \\ a & 0 & 0 & 0 \end{pmatrix}.$$

A curious special case arises in relativity with no counterpart in Euclidean geometry. This is the group of *null rotations*, with

$$L_s = \begin{pmatrix} 1+\frac{1}{2}s^2 & \frac{1}{2}s^2 & s & 0 \\ -\frac{1}{2}s^2 & 1-\frac{1}{2}s^2 & -s & 0 \\ s & s & 1 & 0 \\ 0 & 0 & 0 & 1 \end{pmatrix}, \qquad G = \begin{pmatrix} 0 & 0 & 1 & 0 \\ 0 & 0 & -1 & 0 \\ 1 & 1 & 0 & 0 \\ 0 & 0 & 0 & 0 \end{pmatrix}. \tag{9.10}$$

In this case, we have $\operatorname{tr}(G^2) = 0$. In fact (9.9) and (9.10) exhaust the possibilities, up to conjugation.

Exercise 9.9

Show that with L_s given by (9.10), $L_s L_{s'} = L_{s+s'}$.

Proposition 9.6

For some choice of inertial coordinate system, every one-parameter subgroup of the Lorentz group is given either by (9.9) for some $a, b \in \mathbb{R}$ if $\operatorname{tr}(\gamma^2) \neq 0$ or else by (9.10) if $\operatorname{tr}(\gamma^2) = 0$.

Proof

We can choose the coordinate axes so that the vectors $\boldsymbol{e}$ and $\boldsymbol{b}$ with components in (9.8) lie in the y, z plane. Then $e_1 = b_1 = 0$. If we then make a further change of coordinates, with

$$M = \begin{pmatrix} \cosh\phi & \sinh\phi & 0 & 0 \\ \sinh\phi & \cosh\phi & 0 & 0 \\ 0 & 0 & \cos\theta & -\sin\theta \\ 0 & 0 & \sin\theta & \cos\theta \end{pmatrix}$$

then this condition is preserved, and

$$\begin{pmatrix} e_2 & e_3 \\ b_3 & -b_2 \end{pmatrix} \mapsto \begin{pmatrix} \cosh\phi & -\sinh\phi \\ -\sinh\phi & \cosh\phi \end{pmatrix} \begin{pmatrix} e_2 & e_3 \\ b_3 & -b_2 \end{pmatrix} \begin{pmatrix} \cos\theta & -\sin\theta \\ \sin\theta & \cos\theta \end{pmatrix}.$$

That is $\alpha \mapsto \alpha/w$, $\beta \mapsto w\beta$, where

$$w = \exp(\phi + \mathrm{i}\theta), \quad \alpha = e_2 + b_3 + \mathrm{i}(e_3 - b_2), \quad \beta = e_2 - b_3 - \mathrm{i}(e_3 + b_2).$$

If we choose ϕ and θ so that $w^2 = -\alpha/\beta$, then we shall have $e_2 = b_2 = 0$ in the new matrix of the generator, and G will be of the form (9.9). The exceptional case is that in which $\alpha = 0$ or $\beta = 0$. If necessary by making a rotation through π about the y-axis, one can ensure that the former holds. Then an appropriate choice of ϕ and θ reduces the generator to that in (9.10). □

9.9 Isometries

The symmetries of Minkowski space are the *isometries*—the maps $\iota : \mathbb{M} \to \mathbb{M}$ that preserve its structure as an affine space, and preserve the inner product of four-vectors.

Definition 9.7

A map $\iota : \mathbb{M} \to \mathbb{M}$ is an isometry if there exists $\lambda \in \mathbb{L}$ such that

$$\iota\big(\rho(E, X)\big) = \rho\big(\iota(E), \lambda(X)\big)$$

for every $E \in \mathbb{M}$, $X \in \mathbb{V}$.

If X is the displacement four-vector from E to F, then λX is the displacement four-vector from $\iota(E)$ to $\iota(F)$. Given a choice of basis and an origin, an isometry is given in inertial coordinates by

$$x^a \mapsto \sum_b L^a{}_b x^b + C^a$$

where L is the matrix of λ and $\rho(0, C) = \iota(0)$. The isometries form a group $\mathbb{P}$ (the *Poincaré group*), and the map $\iota \to \lambda$ is a homomorphism onto $\mathbb{L}$.

Given a one-parameter group $\iota_s : \mathbb{M} \to \mathbb{M}$ of isometries, we define a vector field K on $\mathbb{M}$ by

$$K(E) = \left.\frac{\mathrm{d}X_s}{\mathrm{d}s}\right|_{s=0} \qquad \text{where} \quad \rho(E, X_s) = \iota_s(E).$$

Thus for small s, sK gives the infinitesimal displacement from E to $\iota_s(E)$. Such vector fields are called *Killing vectors*. They are characterized by the condition that we met in the context of rigid body motion, that $g(X, \nabla_X K) = 0$ at every event for every four-vector X. That is, K satisfies the *Killing equation*

$$\partial_a K_b + \partial_b K_a = 0.$$

The general solution is

$$K^a = G^a{}_b x^b + T^a$$

where $G^a{}_b$ and T^a are constants, and $G_{ab} = -G_{ba}$. The matrix $G = \big(G^a{}_b\big)$ is the generator of the corresponding one-parameter group of Lorentz transformations.

Exercise 9.10

Show that the Killing equation implies that

$$\begin{aligned} \partial_a\partial_b K_c + \partial_a\partial_c K_b &= 0 \\ \partial_b\partial_c K_a + \partial_b\partial_a K_c &= 0 \\ \partial_c\partial_a K_b + \partial_c\partial_b K_a &= 0 \end{aligned}$$

and hence that $\partial_a\partial_b K_c = 0$. Find the general solution.

The solutions in turn determine the one-parameter subgroups; they form a 10-dimensional vector space—the *Lie algebra* of the isometry group—four from the components of the translation T and six from the generators of the Lorentz group. The Killing vector corresponding to the rotations in the x, y plane has components $(0, -y, x, 0)$; that corresponding to the standard Lorentz transformation has components $(z, t, 0, 0)$.

Exercise 9.11

Suppose that K is a Killing vector and Φ is a four-potential for an electromagnetic field, with the property that

$$K^a\partial_a\Phi_b + \Phi_a\partial_b K^a = 0\,.$$

Show that $K^a(V_a + \kappa\Phi_a)$ is a constant of the motion for a charged particle with four-velocity V, where κ is a constant that should be determined in terms of the charge and rest mass of the particle.

9.10 The Riemann Sphere and Spinors

In Chapter 6, it was shown that a moving sphere always appears spherical when observed visually. This can be seen in another way through an isomorphism between the proper, orthochronous Lorentz group $\mathbb{L}_+$ and the group of Möbius transformations of the Riemann sphere.

The group $\mathbb{L}_+$ acts on the space of future-pointing null vectors K by

$$\begin{pmatrix} \kappa \\ k_1 \\ k_2 \\ k_3 \end{pmatrix} \mapsto L \begin{pmatrix} \kappa \\ k_1 \\ k_2 \\ k_3 \end{pmatrix},$$

where κ and $\boldsymbol{k}$ are the temporal and spatial parts of K. Each K determines a point of the unit sphere by $\boldsymbol{e} = \boldsymbol{k}/\kappa$, since $\kappa^2 = \boldsymbol{k}\,.\,\boldsymbol{k}$ and therefore $\boldsymbol{e}\,.\,\boldsymbol{e} = 1$;

and conversely, each point of the sphere determines a future-pointing null four-vector, uniquely up to multiplication by a positive scalar. So if we write $K \sim K'$ whenever $K = aK'$ for some $a > 0$, then the set of equivalence classes of null future-pointing four-vectors is identified with the unit sphere in $\mathbb{R}^3$. We shall denote the relationship between K and its equivalence class by $\boldsymbol{e} = [K]$. In physical terms, K is tangent to the worldline of a photon and $\boldsymbol{e}$ is the direction in space in which it travels.

Since $L(aK) = aLK$, the proper orthochronous Lorentz group also acts on the unit sphere by $\boldsymbol{e} \mapsto L\boldsymbol{e} = [LK]$. If L relates the inertial frames of two observers and if the first observer sees a photon travelling in the direction $\boldsymbol{e}$, then the second sees it travelling in the direction of $L\boldsymbol{e}$.

The Möbius group, or projective general linear group $\mathrm{PGL}(2, \mathbb{C})$, also acts on the unit sphere. Its action is given by identifying the sphere with the Riemann sphere by stereographic projection. We write $\boldsymbol{e} = (x, y, z)$, with $x^2 + y^2 + z^2 = 1$, and put

$$\zeta(\boldsymbol{e}) = \frac{x + \mathrm{i}y}{1 - z}$$

for $z \neq 1$, and $\zeta(\boldsymbol{e}) = \infty$ when $z = 1$. This labels the points of the sphere by complex numbers, together with $\zeta = \infty$, which is the point $(0, 0, 1)$. The Möbius group M then becomes the group of transformations of the sphere given by

$$\zeta \mapsto \frac{a + b\zeta}{c + d\zeta} \tag{9.11}$$

where $a, b, c, d \in \mathbb{C}$, with $ad - bc \neq 0$. Two such transformations with parameters a, b, c, d and a', b', c', d' are the same whenever

$$\begin{pmatrix} a & b \\ c & d \end{pmatrix} = \mu \begin{pmatrix} a' & b' \\ c' & d' \end{pmatrix}, \qquad \mu \neq 0 \in \mathbb{C}.$$

The group law is given by matrix multiplication; so

$$M = \mathrm{GL}(2, \mathbb{C})/\mathbb{C} = \mathrm{SL}(2, \mathbb{C})/\mathbb{Z}_2\,.$$

Proposition 9.8

$M = \mathbb{L}_+$.

Proof

The isomorphism is defined by identifying the two actions on the Riemann sphere. The rotations

$$\begin{pmatrix} x \\ y \\ z \end{pmatrix} \mapsto \begin{pmatrix} \cos\theta & -\sin\theta & 0 \\ \sin\theta & \cos\theta & 0 \\ 0 & 0 & 1 \end{pmatrix} \begin{pmatrix} x \\ y \\ z \end{pmatrix} \tag{9.12}$$

($\theta \in [0, 2\pi)$) and

$$\begin{pmatrix} x \\ y \\ z \end{pmatrix} \mapsto \begin{pmatrix} z \\ x \\ y \end{pmatrix} \tag{9.13}$$

coincide in their actions with, respectively, the Möbius transformations

$$\zeta \mapsto \mathrm{e}^{\mathrm{i}\theta}\zeta \quad \text{and} \quad \zeta \mapsto \frac{\zeta + \mathrm{i}}{\zeta - \mathrm{i}}. \tag{9.14}$$

The Lorentz transformation

$$L = \begin{pmatrix} \cosh\phi & 0 & 0 & \sinh\phi \\ 0 & 1 & 0 & 0 \\ 0 & 0 & 1 & 0 \\ \sinh\phi & 0 & 0 & \cosh\phi \end{pmatrix} \tag{9.15}$$

($\phi \in \mathbb{R}$) acts on the sphere by

$$L : \begin{pmatrix} x \\ y \\ z \end{pmatrix} \mapsto \frac{1}{\cosh\phi + z\sinh\phi} \begin{pmatrix} x \\ y \\ \sinh\phi + z\cosh\phi \end{pmatrix}$$

and hence by the Möbius transformation $\zeta \to \mathrm{e}^{\phi}\zeta$. Now the rotations (9.12), for different values of θ, together with (9.13), generate SO(3). When we include the Lorentz transformation L (which is the standard Lorentz transformation, but for a permutation of the spatial axes), we generate the whole of the proper orthochronous Lorentz group. Thus we have a group homomorphism $\mathbb{L}_+ \to M$. It is surjective because the Möbius transformations (9.14), together with

$$\zeta \mapsto \mathrm{e}^{\phi}\zeta,$$

generate the whole Möbius group; and it is an isomorphism because only the identity element $I \in \mathbb{L}_+$ gives the identity transformation of the sphere, so the kernel of the homomorphism is $\{I\}$. □

An immediate application is a second, perhaps more fundamental, proof of Penrose's proposition that a moving sphere always appears to have a circular outline, when observed visually [6]. This is now seen to be a simple consequence of the fact that Möbius transformations map circles on the Riemann sphere to circles (see, for example, Priestley's book [7]).

A second consequence is the existence of a two-to-one homomorphism $\mathrm{SL}(2, \mathbb{C}) \to \mathbb{L}_+$, by mapping

$$\begin{pmatrix} a & b \\ c & d \end{pmatrix} \in \mathrm{SL}(2, \mathbb{C})$$

to the Möbius transformation (9.11) ($\mathrm{SL}(2, \mathbb{C})$ is the group of 2×2 complex matrices with unit determinant; the kernel consists of the identity and minus the identity in $\mathrm{SL}(2, \mathbb{C})$.) If we disregard the fact that this is not actually an isomorphism, then we have other candidates for 'relativistic fields' in the linear representations of $\mathrm{SL}(2, \mathbb{C})$. These are (incorrectly) called the *spinor representations* of the Lorentz group; they appear in relativistic quantum mechanics as the wave functions of particles with half-integral spin. The fact that they are not 'proper representations' emerges in the curious phenomenon that such a spinor wave function changes sign under a continuous rotation through 2π.

Appendix A: Notes on Exercises

1.2 (i) With $x' = y' = z' = 0$, we have

$$\begin{pmatrix} x \\ y \\ z \end{pmatrix} = \begin{pmatrix} v_1 \\ v_2 \\ v_3 \end{pmatrix} t' + \begin{pmatrix} c_1 \\ c_2 \\ c_3 \end{pmatrix} = \begin{pmatrix} v_1 \\ v_2 \\ v_3 \end{pmatrix} (t - c_0) + \begin{pmatrix} c_1 \\ c_2 \\ c_3 \end{pmatrix},$$

so the origin of the frame R' moves with velocity (v_1, v_2, v_3) with respect to R.

(ii) Suppose that, in block form,

$$\begin{aligned} \begin{pmatrix} t \\ r \end{pmatrix} &= \begin{pmatrix} 1 & 0 \\ v & H \end{pmatrix} \begin{pmatrix} t' \\ r' \end{pmatrix} + \begin{pmatrix} c_0 \\ c \end{pmatrix} \\ \begin{pmatrix} t' \\ r' \end{pmatrix} &= \begin{pmatrix} 1 & 0 \\ v' & H' \end{pmatrix} \begin{pmatrix} t'' \\ r'' \end{pmatrix} + \begin{pmatrix} c_0' \\ c' \end{pmatrix}, \end{aligned}$$

where r is the column vector with entries x, y, z, and so on. Then

$$\begin{pmatrix} t \\ r \end{pmatrix} = \begin{pmatrix} 1 & 0 \\ v + Hv' & HH' \end{pmatrix} \begin{pmatrix} t'' \\ r'' \end{pmatrix} + \begin{pmatrix} c_0 + c_0' \\ c + c_0' v + Hc' \end{pmatrix}.$$

The velocity $v + Hv'$ of the composite transformation is the vector sum of the velocities of R' relative to R (components v_1, v_2, v_3) and the velocity of R'' relative to R (components v_1'', v_2'', v_3'').

1.3 Pappus' theorem is the following. Let A_1, A_2, A_3, B_1, B_2, B_3 be distinct points of the plane. Let C_1 be the point of intersection of the lines A_2B_3 and A_3B_2, C_2 be the point of intersection of the lines A_3B_1 and A_1B_3, and C_3 be the point of intersection of the lines A_1B_2 and A_2B_1. If A_1, A_2, A_3 are collinear and B_1, B_2, B_3 are collinear, then C_1, C_2, C_3 are collinear.

To apply it to the problem at hand, take A_1, A_2, A_3 to be events on the worldline of the Greek army and B_1, B_2, B_3 to be events on the worldline of the Persian army.

3.3 We have

$$\phi(\boldsymbol{r}) = \int_{\boldsymbol{r}' \in V} \frac{k\rho(\boldsymbol{r}')}{|\boldsymbol{r} - \boldsymbol{r}'|}\, \mathrm{d}V' .$$

In the integral, $\boldsymbol{r}$ (the position of the point at which ϕ is evaluated) is fixed, and the integration is over the positions $\boldsymbol{r}'$ of the volume elements in V. In spite of the singularity at $\boldsymbol{r} = \boldsymbol{r}'$, the integral exists.

The change in ϕ induced by moving from $\boldsymbol{r}$ to $\boldsymbol{r} + \boldsymbol{h}$ is the same as the change induced by translating the whole distribution of matter through $-\boldsymbol{h}$. Therefore

$$\operatorname{grad} \phi = \int_{\boldsymbol{r}' \in V} \frac{k \operatorname{grad}' \rho(\boldsymbol{r}')}{|\boldsymbol{r} - \boldsymbol{r}'|}\, \mathrm{d}V' .$$

(ii) Fix $\boldsymbol{r}'$ and introduce spherical polar coordinates with $\boldsymbol{r}'$ as origin. Then $|\boldsymbol{r} - \boldsymbol{r}'| = r$. Now use the fact that if f is a function of r alone, then

$$\nabla^2 f = \frac{1}{r}\frac{\mathrm{d}^2}{\mathrm{d}r^2}\left(rf\right) .$$

(iii) We can find $\nabla^2\phi$ by taking the divergence by differentiating under the integral sign. To avoid the problem of the singularity in the integral, we replace V by V', obtained by excluding a small ball B_ϵ of radius ϵ centred on $\boldsymbol{r}$, and then taking the limit as $\epsilon \rightarrow 0$. The result is

$$\begin{aligned} \nabla^2\phi &= \lim_{\epsilon\rightarrow 0}\int_{V'} k \operatorname{grad}\left(\frac{1}{|\boldsymbol{r} - \boldsymbol{r}'|}\right) \cdot \operatorname{grad}' \rho\, \mathrm{d}V' \\ &= -\lim_{\epsilon\rightarrow 0}\int_{V'} k \operatorname{grad}'\left(\frac{1}{|\boldsymbol{r} - \boldsymbol{r}'|}\right) \cdot \operatorname{grad}' \rho\, \mathrm{d}V' \\ &= -\lim_{\epsilon\rightarrow 0}\int_{S_\epsilon} k\rho \operatorname{grad}'\left(\frac{1}{|\boldsymbol{r} - \boldsymbol{r}'|}\right) . \mathrm{d}\boldsymbol{S}' , \end{aligned}$$

where S_ϵ is the boundary of B_ϵ, by applying the divergence theorem in V'. Now use

$$\operatorname{grad}'\left(\frac{1}{|\boldsymbol{r} - \boldsymbol{r}'|}\right) = -\frac{\boldsymbol{r}' - \boldsymbol{r}}{|\boldsymbol{r} - \boldsymbol{r}'|^3}, \quad \mathrm{d}\boldsymbol{S}' = -\frac{\boldsymbol{r}' - \boldsymbol{r}}{|\boldsymbol{r} - \boldsymbol{r}'|}\, \mathrm{d}S' .$$

On S_ϵ, we have $|\boldsymbol{r}' - \boldsymbol{r}| = \epsilon$. Therefore

$$\nabla^2\phi = -\lim_{\epsilon\rightarrow 0}\int_{S_\epsilon} \frac{k\rho(\boldsymbol{r}')}{\epsilon^2}\, \mathrm{d}S' \rightarrow -4\pi k\rho(\boldsymbol{r}) .$$

3.4 Note that

$$\begin{aligned}
\operatorname{div} \boldsymbol{E} &= -\mathrm{i} \operatorname{grad} \Omega \,.\, \boldsymbol{a}\, \mathrm{e}^{-\mathrm{i}\Omega} = \frac{\mathrm{i}\omega \boldsymbol{e}\,.\,\boldsymbol{a}}{c}\, \mathrm{e}^{-\mathrm{i}\Omega} \\
\operatorname{curl} \boldsymbol{E} &= -\mathrm{i} \operatorname{grad} \Omega \wedge \boldsymbol{a}\, \mathrm{e}^{-\mathrm{i}\Omega} = \frac{\mathrm{i}\omega \boldsymbol{e} \wedge \boldsymbol{a}}{c}\, \mathrm{e}^{-\mathrm{i}\Omega} \\
\frac{\partial \boldsymbol{E}}{\partial t} &= -\mathrm{i} \frac{\partial \Omega}{\partial t} \boldsymbol{a}\, \mathrm{e}^{-\mathrm{i}\Omega} = -\mathrm{i}\omega \boldsymbol{a}\, \mathrm{e}^{-\mathrm{i}\Omega} ,
\end{aligned}$$

and so on. The conditions are: $\boldsymbol{e}$ is a unit vector, $\boldsymbol{e}.\boldsymbol{a} = 0$, and $\boldsymbol{b} = \boldsymbol{e} \wedge \boldsymbol{a}$. The wave is linearly polarized if $\boldsymbol{a}$ is the product of a real vector and a complex scalar. It has circular polarization if

$$\boldsymbol{a}\,.\,\boldsymbol{a} = 0 .$$

3.5 Let S denote the unit sphere. If $\boldsymbol{B} = \operatorname{curl} \boldsymbol{A}$, then

$$\int_S \boldsymbol{B}\,.\,\mathrm{d}\boldsymbol{S} = \int_S \operatorname{curl} \boldsymbol{A}\,.\,\mathrm{d}\boldsymbol{S} = 0$$

by Stokes' theorem. On the other hand,

$$\int_S \frac{\boldsymbol{r}}{r^3}\,.\,\mathrm{d}\boldsymbol{S} = \int_S r^{-2}\, \mathrm{d}S = 4\pi .$$

So $\boldsymbol{r}/r$ is not the curl of a vector field in spite of the fact that its divergence vanishes.

3.6 If we take $\boldsymbol{\alpha} = \boldsymbol{k}$, then $\boldsymbol{\alpha} \wedge \boldsymbol{r} = y\boldsymbol{i} - x\boldsymbol{j}$. Therefore in this case

$$\begin{aligned}
\nabla^2(F\boldsymbol{\alpha} \wedge \boldsymbol{r}) &= \nabla^2(Fy)\boldsymbol{i} - \nabla^2(Fx)\boldsymbol{j} \\
&= \nabla^2(F)(y\boldsymbol{i} - x\boldsymbol{j}) + 2\frac{\partial F}{\partial y}\boldsymbol{i} - 2\frac{\partial F}{\partial x}\boldsymbol{j} \\
&= \nabla^2(F)\boldsymbol{\alpha} \wedge \boldsymbol{r} + 2\boldsymbol{\alpha} \wedge \operatorname{grad} F .
\end{aligned}$$

By linearity, this also holds if $\boldsymbol{\alpha}$ is a constant multiple of $\boldsymbol{k}$; by rotating the axes, it therefore holds for any constant $\boldsymbol{\alpha}$.

If $\boldsymbol{b} = r^{-2} g(kr)\boldsymbol{\alpha} \wedge \boldsymbol{r}$, then $\operatorname{div} \boldsymbol{b} = 0$ and

$$\nabla^2(\boldsymbol{b}) = \frac{k^2}{r^2}\left(g'' - \frac{2g}{k^2 r^2}\right) \boldsymbol{\alpha} \wedge \boldsymbol{r} .$$

Consider $g(x) = \cos x - \sin x / x$.

4.1 (i) By the chain rule,

$$\frac{1}{c}\frac{\partial}{\partial t'} = \gamma(u)\frac{1}{c}\frac{\partial}{\partial t} + \gamma(u)\frac{u}{c}\frac{\partial}{\partial x}, \qquad \frac{\partial}{\partial x'} = \gamma(u)\frac{u}{c^2}\frac{\partial}{\partial t} + \gamma(u)\frac{\partial}{\partial x}.$$

The wave is harmonic in the new coordinates. Its amplitude is unchanged, but its frequency is

$$\omega' = \omega\sqrt{\frac{c-u}{c+u}}.$$

(ii) With $t' = 0$, $x' = -D$, we have

$$t = -\frac{\gamma(u)uD}{c^2} = -\frac{uD}{c^2} + O(u^2/c^2), \qquad x = -\gamma(u)D = -D + O(u^2/c^2).$$

(iii) In this example, the speeds in miles per second are $u = 5/3600$ and $c = 186,000$. Therefore $u/c = 7.5 \times 10^{-9}$. We also have $D/c = 2.2 \times 10^6$ years. Therefore the time is

$$2.2 \times 7.5 \times 10^{-3} \text{ years} = 6 \text{ days}.$$

4.2 For the 'if' part, put

$$L = \begin{pmatrix} p & q \\ r & s \end{pmatrix}.$$

Then the conditions are (i) $p > 0$, (ii) $ps - qr > 0$, (iii) $p^2 - r^2 = 1$, (iv) $pq - rs = 0$, and (v) $q^2 - s^2 = -1$. By (v), we have $s \neq 0$. Put $u/c = r/p$. Show that $|u| < c$ and that

$$L = \gamma(u)\begin{pmatrix} 1 & u/c \\ u/c & 1 \end{pmatrix}.$$

4.3 In the space–time diagram (Figure A.1), the events F and E are simultaneous in the frame of the athlete; E is the event at which the end Q strikes the wall. The key is that the athlete measures the distance between F and E when he measures the length of the pole. The event F is outside the room.

The worldline of a photon emitted at E reaches P at the event G. Since the shock cannot travel along the pole faster than light, the shock does not reach P before G. So the answer is (ii).

5.4 By putting $x' = y' = z' = 0$, we have

$$t = L^0{}_0 t', \quad x = cL^1{}_0 t', \quad y = cL^2{}_0 t', \quad z = cL^3{}_0 t'.$$

Hence,

$$u_1 = \frac{\mathrm{d}x}{\mathrm{d}t} = cL^1{}_0 / L^0{}_0, \qquad \text{etc.}$$

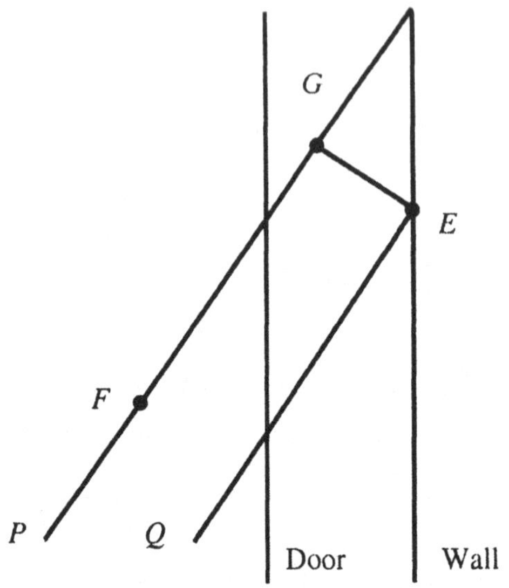

Figure A.1 The athlete and the pole

5.5 All are Lorentz transformation matrices. Only the first and last are proper and orthochronous.

5.6 (i) Rotate the spatial axes so that V has components $(p, q, 0, 0)$, with $p > 0$, $|q| < p$. The problem is then to find u so that $|u| < c$ and

$$\gamma(u)\begin{pmatrix} 1 & u/c \\ u/c & 1 \end{pmatrix}\begin{pmatrix} p \\ q \end{pmatrix} = \begin{pmatrix} a \\ 0 \end{pmatrix}$$

for some $a > 0$. This is solved by $u = -qc/p$. We then have

$$a = \gamma(p^2 - q^2)/p > 0$$

and $g(V, V) = a^2$ (since $g(V, V)$ is invariant).

5.7 (i) This is obvious in an inertial coordinate system chosen so that one of the four-vectors has components $(a, 0, 0, 0)$, where $a > 0$. Another method is to take the four-vectors to be $(U^0, \boldsymbol{u})$, $(V^0, \boldsymbol{v})$, where $U^0 > 0$, $V^0 > 0$, $\boldsymbol{u} \cdot \boldsymbol{u} < (U^0)^2$, $\boldsymbol{v} \cdot \boldsymbol{v} < (V^0)^2$, and to show that

$$(\boldsymbol{u} + \boldsymbol{v}) \cdot (\boldsymbol{u} + \boldsymbol{v}) < (U^0 + V^0)^2 .$$

(ii) The same argument works, with now $|\boldsymbol{u}| = U^0$, $|\boldsymbol{v}| = V^0$, and equality in the last line. The sum is null if and only if $\boldsymbol{u} \cdot \boldsymbol{v} = |\boldsymbol{u}|\,|\boldsymbol{v}|$. That is, if and only if the two four-vectors are proportional.

(iii) Choose the inertial coordinates so that the timelike vector has components $(a, 0, 0, 0)$, with $a \neq 0$.

5.8 Write $X = (X^0, \boldsymbol{x})$, $Y = (Y^0, \boldsymbol{y})$. By choosing the inertial coordinates so that $\boldsymbol{x} = 0$, show as a first step that

$$g(X,Y) > \sqrt{g(X,X)g(Y,Y)}\,.$$

Equality holds when X and Y are proportional. The analogous result in Euclidean geometry is the triangle inequality.

5.9 Show that

$$\operatorname{Grad} g(X,X) = \operatorname{Grad}(c^2t^2 - x^2 - y^2 - z^2) = 2X\,.$$

5.10 By translation, u is a solution for real A. By analytic continuation, it is also a solution for complex A; so are its real and imaginary parts. Take A to have components $(ic\tau, 0, 0, 0)$ for some real constant $\tau > 0$ and show that

$$\operatorname{Im}(u) = \frac{2t\tau c^2}{\left(x^2 + y^2 + z^2 + c^2(\tau^2 - t^2)\right)^2 + 4t^2\tau^2c^4}\,,$$

which is nonsingular since the denominator is everywhere positive. To show that it is bounded, show that

$$c^2\left|\operatorname{Im}(u)\right| \le \frac{2R\tau}{\tau^4 + 2\tau^2R^2}\,,$$

where $R^2 = a^2 + t^2 = (x^2 + y^2 + z^2 + c^2t^2)/c^2$. A nonsingular solution ϕ of Laplace's equation in four dimensions with $\phi = O(R^{-1})$ as $R \to \infty$ must be identically zero (by the maximum principle).

5.11 A harmonic wave is a solution of the wave equation $\Box u = 0$ of the form

$$u = \alpha \cos(\Omega + \epsilon)\,,$$

where $\Omega = \omega(ct - \boldsymbol{e}\,.\,\boldsymbol{r})/c$, for some constant $\alpha, \omega, \epsilon, \boldsymbol{e}$, with $\omega > 0$ (the frequency) and $\boldsymbol{e}\,.\boldsymbol{e} = 1$.

If we put $K = c\nabla\Omega$, then K has components $\omega(1, \boldsymbol{e})$, and is therefore a future-pointing null vector. In any other coordinate system, Ω will again be a linear function of the coordinates, with $K^0 = \partial_t\Omega$, $K^1 = -c\partial_x\Omega$ and so on. So we shall have

$$\Omega = c^{-1}(K^0ct - K^1x - K^2y - K^3z) + \text{constant}\,.$$

Thus in these coordinates, u again has the form of a harmonic wave with frequency K^0. If we choose the coordinates so that the observer is at rest, then V has components $(c, 0, 0, 0, 0)$ and we have

$$cK^0 = g(K, V)\,.$$

So the observed frequency is $g(K, V)/c$. The waves appear to an observer with general four-velocity V to have the same frequency if

$$\begin{aligned} \sqrt{3}V^0 - V^1 - V^2 - V^3 &= p\lambda \\ \sqrt{3}V^0 - V^1 + V^2 + V^3 &= q\lambda \\ \sqrt{3}V^0 + V^1 - V^2 + V^3 &= r\lambda \\ \sqrt{3}V^0 + V^1 + V^2 - V^3 &= s\lambda\,, \end{aligned}$$

for some $\lambda \in \mathbb{R}$, with $g(V, V) = c^2$.

6.3 This is essentially the same result as in in Exercise 5.8, but in a concrete setting. Since σU is the four-vector from A to B, τV is the four-vector from B to C, and $\tau' V'$ is the four-vector from A to C, we have

$$\tau' V' = \sigma U + \tau V\,.$$

By taking the inner product of each side with itself, and using the fact that $g(U, U) = c^2$, and so on, we have

$$c^2\tau'^2 = c^2\sigma^2 + c^2\tau^2 + 2g(U, V)\sigma\tau\,.$$

Now use $\gamma(v) = 1/\sqrt{1 - v^2/c^2} > 1$. In classical physics, $\tau' = \sigma + \tau$.

6.4 From the definition, A has components

$$(c \sinh\phi, c\cosh\phi, 0, 0)\frac{\mathrm{d}\phi}{\mathrm{d}\tau}\,.$$

6.5 (i) The vector Z has components

$$\frac{2c^2}{a}\Big(\sinh(a\tau/c), \cosh(a\tau/c)\Big)\,.$$

Therefore, $g(Z, Z) = -4c^4/a^2$.

(ii) The four-velocity of the first rocket at the event A has components

$$c\big(\cosh(a\tau/c), -\sinh(a\tau/c)\big)\,.$$

An observer reckons that two events are simultaneous if and only if the vector between them is orthogonal to his four-velocity. The resolution of the apparent paradox is that A gets earlier on the first worldline as B gets later on the second.

7.1 The four-momenta of the particles are $P = (E/c, \boldsymbol{p})$ and $Q = m(c, 0)$ in the given inertial frame. In the centre-of-mass frame, the four-velocity V of the centre of mass decomposes as $(c, 0)$ and $P+Q$ decomposes as $(E'/c, 0)$. Moreover, $g(V, V) = c^2$. Therefore

$$V = \frac{c(P+Q)}{\sqrt{g(P+Q, P+Q)}}, \qquad E' = g(V, P+Q).$$

Now use $E'^2 = c^2 g(P+Q, P+Q)$.

7.3 (i) The four-momentum equation is

$$mU = m'U' + MV,$$

where $U = \gamma(u)(c, u)$, $U' = \gamma(u')(c, u')$, $V = \gamma(v)(c, -v)$ in the given frame. Now $g(U, U) = g(V, V) = g(U', U') = c^2$ and $g(U, V) = \gamma(w)c^2$. Hence

$$\begin{aligned} m'^2c^2 &= g(m'U', m'U') \\ &= g(mU - MV, mU - MV) \\ &= m^2c^2 + M^2c^2 - 2mMc^2\gamma(w). \end{aligned}$$

By taking the inner product of the four-momentum equation with (u, c) (which is orthogonal to U), we have

$$0 = m'\gamma(u')c(u - u') + M\gamma(v)c(u + v).$$

(ii) This is a direct application of the transformation rule, using the fact that

$$\begin{pmatrix} ct \\ x \end{pmatrix} = \gamma(u) \begin{pmatrix} 1 & u/c \\ u/c & 1 \end{pmatrix} \begin{pmatrix} ct' \\ x' \end{pmatrix}$$

is the Lorentz transformation between the coordinates of the given frame, (t, x), and those in which the rocket is at rest before the ejection, (t', x').

(iii) It follows from (ii) that

$$\gamma(v)(u + v) = \gamma(u)\gamma(w)w(1 - u^2/c^2) = w\gamma(w)/\gamma(u).$$

Hence, by the first equation in (i),

$$m'\gamma(u')(u - u') + Mw\gamma(w)/\gamma(u) = 0$$

and the result follows from the second equation in (i).

(iv) By putting $m' = m + \delta m$, $u' = u + \delta u$, we have

$$mm'\gamma(u)\gamma(u')\delta u = \tfrac{1}{2}w(-2m\delta m - \delta m^2 + M^2)\,.$$

As $M \to 0$, we have $u' \to u$ and $m' \to m$. Hence

$$m\gamma(u)^2 \frac{\mathrm{d}u}{\mathrm{d}m} + w = 0\,,$$

giving the required result.

(v) It follows that

$$\begin{aligned} -\int \frac{w\,\mathrm{d}m}{m} &= \int \frac{\mathrm{d}u}{1 - u^2/c^2} \\ &= \int \frac{c}{2}\left(\frac{1}{c-u} + \frac{1}{c+u}\right)\mathrm{d}u \\ &= \log\left(\frac{c+u}{c-u}\right)^{c/2}\,. \end{aligned}$$

The first equality follows. The second follows from $u = c\tanh(a\tau/c)$.

In the last part, we can take (very roughly) $c/a = c/g = 1$ year. The round trip time is about 40 years (measured in the rocket). Hence

$$\frac{m_{\text{initial}}}{m_{\text{final}}} = \exp(40 \times 186{,}000 \times 3600/1000)\,.$$

Thus 10^{10^7} tons of fuel are required for each ton of payload returned to earth. This suggests that an investment in such an expedition would not return a good yield.

8.1 Note that under both rotations and standard Lorentz transformations, the transformation rule for $\boldsymbol{E}$ and $\boldsymbol{B}$ is unchanged under $\boldsymbol{E} \mapsto -c\boldsymbol{B}$, $c\boldsymbol{B} \mapsto \boldsymbol{E}$.

Show that $\operatorname{tr}(gF^*gF) = -4c\boldsymbol{E}\,.\,\boldsymbol{B}$. Establish invariance by using the cyclic symmtery of the trace.

8.3 The first part follows from

$$\operatorname{Div}\Phi = -c^{-1}g(P,K)\sin\Omega, \qquad \Box\Phi = -c^{-2}g(K,K)P\cos\Omega\,.$$

Suppose that K is null and P is orthogonal to K. Put $K = \omega(1,\boldsymbol{e})$, $P = (\boldsymbol{p}\,.\,\boldsymbol{e},\boldsymbol{p})$, $X = (ct,\boldsymbol{r})$. Then

$$\Phi = (\phi, c\boldsymbol{A}) = (\boldsymbol{p}\,.\,\boldsymbol{e},\boldsymbol{p})\cos(\omega t - \omega\boldsymbol{e}\,.\,\boldsymbol{r}/c)\,.$$

Therefore

$$\begin{aligned} c\boldsymbol{B} &= c\,\mathrm{curl}\,\boldsymbol{A} = c^{-1}\omega \boldsymbol{e} \wedge \boldsymbol{p} \sin(\Omega) \\ \boldsymbol{E} &= -\frac{\partial \boldsymbol{A}}{\partial t} - \nabla\phi = c^{-1}\omega\big(\boldsymbol{p} - (\boldsymbol{p}\,.\,\boldsymbol{e})\boldsymbol{e}\big) \sin\Omega\,. \end{aligned}$$

This is a linearly polarized monochromatic plane wave.

9.1 It may help to note that if T_1 and T_2 are timelike and if $g(T_1, T_2) > 0$, then $sT_1 + (1 - s)T_2$ is timelike for all $s \in [0, 1]$.

Appendix B: Vector Calculus

Three-vector Identities

The *gradient* of a function f is defined by

$$\operatorname{grad} f = \partial_x f \boldsymbol{i} + \partial_y f \boldsymbol{j} + \partial_z f \boldsymbol{k}\,,$$

where $\partial_x = \partial/\partial x$, and so on. The curl and divergence of a vector field $\boldsymbol{v} = a\boldsymbol{i} + b\boldsymbol{j} + c\boldsymbol{k}$ are defined, respectively, by

$$\operatorname{div} \boldsymbol{v} = \partial_x a + \partial_y b + \partial_z c \tag{B.1}$$

$$\operatorname{curl} \boldsymbol{v} = (\partial_y c - \partial_z b)\boldsymbol{i} + (\partial_z a - \partial_x c)\boldsymbol{j} + (\partial_x b - \partial_y a)\boldsymbol{k}\,. \tag{B.2}$$

For any vector fields $\boldsymbol{u}, \boldsymbol{v}, \boldsymbol{w}$ and function f

$$\boldsymbol{u} \,.\, (\boldsymbol{v} \wedge \boldsymbol{w}) = \boldsymbol{w} \,.\, (\boldsymbol{u} \wedge \boldsymbol{v}) \tag{B.3}$$

$$\boldsymbol{u} \wedge (\boldsymbol{v} \wedge \boldsymbol{w}) = (\boldsymbol{u} \,.\, \boldsymbol{w})\boldsymbol{v} - (\boldsymbol{u} \,.\, \boldsymbol{v})\boldsymbol{w} \tag{B.4}$$

$$\operatorname{curl}(\operatorname{grad} f) = 0 \tag{B.5}$$

$$\operatorname{div}(\operatorname{curl} \boldsymbol{u}) = 0 \tag{B.6}$$

$$\operatorname{curl}(\operatorname{curl} \boldsymbol{u}) = \operatorname{grad}(\operatorname{div} \boldsymbol{u}) - \nabla^2 \boldsymbol{u} \tag{B.7}$$

$$\operatorname{div}(f\boldsymbol{u}) = \operatorname{grad} f \,.\, \boldsymbol{u} + f \operatorname{div} \boldsymbol{u} \tag{B.8}$$

$$\operatorname{curl}(f\boldsymbol{u}) = \operatorname{grad} f \wedge \boldsymbol{u} + f \operatorname{curl} \boldsymbol{u} \tag{B.9}$$

$$\operatorname{div}(\boldsymbol{u} \wedge \boldsymbol{v}) = \boldsymbol{v} \,.\, \operatorname{curl} \boldsymbol{u} - \boldsymbol{u} \,.\, \operatorname{curl} \boldsymbol{v} \tag{B.10}$$

$$\operatorname{curl}(\boldsymbol{u} \wedge \boldsymbol{v}) = (\operatorname{div} \boldsymbol{v})\boldsymbol{u} - (\operatorname{div} \boldsymbol{u})\boldsymbol{v} + (\boldsymbol{v} \,.\, \boldsymbol{\nabla})\boldsymbol{u} - (\boldsymbol{u} \,.\, \boldsymbol{\nabla})\boldsymbol{v}\,. \tag{B.11}$$

Four-vector Identities

The *four-gradient* Grad f of a function f on space–time is the four-vector field with components

$$\left(\frac{1}{c}\frac{\partial f}{\partial t}, -\frac{\partial f}{\partial x}, -\frac{\partial f}{\partial y}, -\frac{\partial f}{\partial z}\right).$$

The *four-divergence* of a four-vector field V is the scalar function

$$\operatorname{Div} V = \frac{1}{c}\frac{\partial V^0}{\partial t} + \frac{\partial V^1}{\partial x} + \frac{\partial V^2}{\partial y} + \frac{\partial V^3}{\partial z}.$$

For any function u and four-vector field V

$$\operatorname{Div}(uV) = u \operatorname{Div} V + g(V, \operatorname{Grad} u) \tag{B.12}$$

$$\operatorname{Div}(\operatorname{Grad} u) = \Box u, \tag{B.13}$$

where $\Box$ is the d'Alembertian, defined by

$$\Box u = \frac{1}{c^2}\frac{\partial^2 u}{\partial t^2} - \frac{\partial^2 u}{\partial x^2} - \frac{\partial^2 u}{\partial y^2} - \frac{\partial^2 u}{\partial z^2}.$$

Existence of Potentials

In the following, the precise conditions on the functions involved have not been specified explicitly. It is certainly sufficient that they should be *smooth* (infinitely differentiable), but the results also hold under much weaker conditions.

Proposition B.1

Suppose that a and b are given functions of x, y, z on some open ball in $\mathbb{R}^3$. Then there exists a function $\psi(x, y, z)$ such that

$$\frac{\partial \psi}{\partial x} = a, \qquad \frac{\partial \psi}{\partial y} = b,$$

if and only if $\partial_x b = \partial_y a$.

Proof

The 'only if' part is immediate. To prove the 'if' part, suppose that $a_y = b_x$, where the subscripts denote partial derivatives; and suppose, without loss of generality, that the ball contains the origin. Define $\psi(x, y, z)$ in the ball by

$$\psi(x, y, z) = \int_0^x a(t, y, z)\,\mathrm{d}t + \int_0^y b(0, t, z)\,\mathrm{d}t.$$

Then $\psi_x = a$ and

$$\begin{aligned}\psi_y(x,y,z) &= b(0,y,z) + \int_0^x a_y(t,y,z)\,\mathrm{d}t \\ &= b(0,y,z) + \int_0^x b_t(t,y,z)\,\mathrm{d}t \\ &= b(x,y,z)\,.\end{aligned}$$

□

Corollary B.2

Suppose that $\operatorname{curl}\boldsymbol{u} = 0$ is some open ball. Then there exists a function $\phi(x,y,z)$ such that $\boldsymbol{u} = \operatorname{grad}\phi$.

Proof

Write $\boldsymbol{u} = a\boldsymbol{i} + b\boldsymbol{j} + c\boldsymbol{k}$. Since $a_y = b_x$, there exists ψ such that $a = \psi_x$, $b = \psi_y$. Put $\boldsymbol{v} = \boldsymbol{u} - \operatorname{grad}\psi$. Then $\operatorname{curl}\boldsymbol{v} = 0$, by (B.5), and $\boldsymbol{v} = h\boldsymbol{k}$ for some function h since the $\boldsymbol{i}$ and $\boldsymbol{j}$ components of $\boldsymbol{v}$ vanish. But $\operatorname{curl}\boldsymbol{v} = 0$ implies $h_x = h_y = 0$, and so $h = h(z)$ is a function of z alone. The proof is now completed by putting

$$\phi(x,y,z) = \psi(x,y,z) + \int_0^z h(t)\,\mathrm{d}t\,.$$

□

Corollary B.3

Let $\boldsymbol{u}$ be a vector field on an open ball such that $\operatorname{div}\boldsymbol{u} = 0$. Then there exists a vector field $\boldsymbol{w}$ such that $\boldsymbol{u} = \operatorname{curl}\boldsymbol{w}$.

Proof

Write $\boldsymbol{u} = a\boldsymbol{i} + b\boldsymbol{j} + c\boldsymbol{k}$. If $c = 0$, then $a_x = -b_y$ and we can take $\boldsymbol{w} = \psi\boldsymbol{k}$ where

$$\psi_y = a, \qquad \psi_x = -b\,.$$

In general, we can reduce to this case by putting $\boldsymbol{v} = \boldsymbol{u} - \operatorname{curl}(\phi\boldsymbol{i})$, where ϕ is a function of x, y, z chosen so that $\phi_y = c$. Then

$$\operatorname{div}\boldsymbol{v} = 0 \qquad \text{and} \qquad \boldsymbol{v}\,.\,\boldsymbol{k} = 0\,,$$

so $\boldsymbol{v} = \operatorname{curl}\boldsymbol{x}$ for some $\boldsymbol{x}$ and $\boldsymbol{u} = \operatorname{curl}(\boldsymbol{x} + \phi\boldsymbol{i})$. □

Bibliography

[1] H. Bondi *Assumption and myth in physical theory.* Cambridge University Press, Cambridge, 1967.

[2] A. Einstein *On the electrodynamics of moving bodies.* A translation of the paper can be found in *The principle of relativity* by H. A. Lorentz, A. Einstein, H. Minkowski, and H. Weyl, with notes by A. Sommerfeld. Dover, New York, 1952.

[3] Galileo *Dialogue concerning the two chief world systems: The Ptolemaic and Copernican.* 2nd Revised edition. Edited translated by S. Drake. University of California Press, Berkeley, California, 1967.

[4] S. Kobayashi and K. Nomizu *Foundations of differential geometry*, Volume I. Wiley, New York, 1963.

[5] E. A. Milne *Relativity, gravitation and world-structure.* Oxford University Press, Oxford, 1935.

[6] R. Penrose *The apparent shape of a relativistically moving sphere. Proc. Camb. Phil. Soc.* **55**, 137–9 (1959).

[7] H. A. Priestley *Introduction to complex analysis.* Oxford University Press, Oxford, 1990.

[8] W. Rindler *Introduction to special relativity.* Oxford University Press, Oxford, 1991.

[9] W. Rindler *Essential relativity.* Springer-Verlag, Berlin, 1960.

[10] J. L. Synge *Relativity, the special theory.* North-Holland, Amsterdam, 1955.

[11] I. Tolstoy in *James Clerk Maxwell, a biography.* Canongate, Edinburgh, 1981.

[12] A. Trautman, F. A. E. Pirani, and H. Bondi *Lectures on general relativity.* Eds. S. Deser and K. W. Ford. Prentice–Hall, Englewood Cliffs, New Jersey, 1965.

[13] E. P. Wigner *The unreasonable effectiveness of mathematics. Commun. Pure Appl. Math.* **13**, 1–14 (1960).

Index